The Original American Wheat Dolly

Text and Artwork by

Mary Lomen

Dedication

To my family whose wonderful disposition and encouraging optimism have created the perfect climate for enjoying each day to the fullest!

I owe this book to them.

And especially to my son, Bill, for all his good natured help and support.

International Standard Book Number: 0-942323-15-7

Published by
North American Heritage Press
A DIVISION OF
CREATIVE MEDIA, INC.
P.O. Box 1
Minot, North Dakota 58702
701/852-5552

Printed in the United States of America

"Wheat signifies life itself."

Table of Contents

Table of Contents *(continued)*

Table of Contents *(continued)*

Table of Contents *(continued)*

Introduction

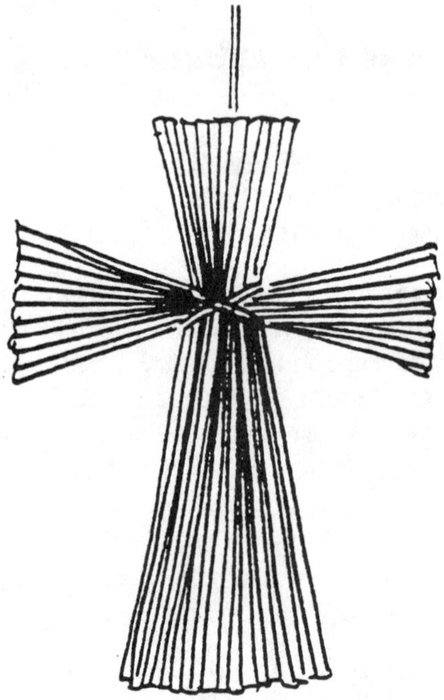

THE WHEAT DOLLY
"Fruit Of The Harvest"

Illuminating the home with wheat dollies* is a delightful age-old custom which is rapidly becoming a popular home decorating motif in the United States.

These beautiful, all-natural, hand-crafted ornaments have been used for countless centuries, and they are found in homes throughout the world. As living works of art they are especially esteemed by many Europeans, who traditionally decorate their homes with these symbols of health, happiness, and good fortune.

Because of the Wheat Dollies all-natural material and its charming and quaint appearance, I have found that they create instant interest.

Wheat is one of our oldest known foods. It is almost complete in itself and is so valuable nutritionally that it has special meaning for almost everyone throughout the world.

We use the Wheat Dolly as a decorative symbol of peace, love, prosperity, thankfulness, and hope in the home throughout the year. Because of its spiritual and nutritional significance, as well as its beauty, we consider wheat one of the most blessed, God-given gifts we have.

You may do no more than get acquainted with Wheat Dollies or become a collector of these charming ornaments, or you may discover, as I have, the enjoyment of creating with everyday materials found around you. This is one of the main reasons I've written this book.

Mary Lonnen

**Wheat Dollies were traditionally known as "Corn Dollies." Here the word "corn" means all grain: wheat, barley, oats, rye, etc.*

Wheat Dollies, How They Were Named

Halsten, my husband, and I discovered the English Corn Dolly for the first time in 1971. We were traveling in England and we had just come down to breakfast at a charming country hotel; the first thing that caught our eye were two large lovely hand woven Corn Dollies hanging on the wall near our table. I could not imagine anything so beautiful made entirely of wheat. I was later able to purchase two small Corn Dollies. On the back of each package were instructions on how to make them.

At that time working with wheat was not new to me as we own a large wheat farm in Wildrose, North Dakota and I had been selling floral wheat for several years to large floral supply companies. But I never dreamed of finding anything as delightful as these ornaments, made of wheat.

Returning to the United States, my daughter, Virginia and I began to experiment with wheat weaving. We copied the two dollies I had purchased in England. We also designed some of our own. For want of instructions we were able to make ten "original dollies simply by twisting, tieing, and using the methods of weaving given us on the Corn Dolly packages. With these few samples, our small Wheat Dolly Company was formed.

My first drawings with written instructions were for the Wheat Dolly workers to follow. These women were very interested and enthusiastic in this new endeavor.

Information on wheat weaving was very limited, even in libraries and book stores. My son, Bill, was in Australia at the time and he found a Corn Dolly book, published in England. He sent it to us which proved to be very helpful.

Traditions

In many countries around the world entire families work together growing and weaving wheat. In the process they keep this tradition alive and pass this art of weaving wheat from one generation to the next. Throughout Europe there are special wheat harvest festivals, and the custom of wheat weaving is continually renewed, keeping this delightful tradition alive and growing.

I receive many letters from people around the United States, who have become interested in learning more about this fascinating art. Others want to purchase or obtain instructions on certain patterns or designs. Many wheat weaving clubs have been formed and members have a chance to learn what others are doing. They meet with one another, and send monthly newsletters which generate further interest. It is hoped that this book will not only help to create interest and preserve the tradition of the Wheat Dolly, but also encourage others to pursue the art of Wheat Weaving.

North Dakota Treasure - An Original Design by Mary Lomen

The Ancestry Of Wheat

CHAPTER 1

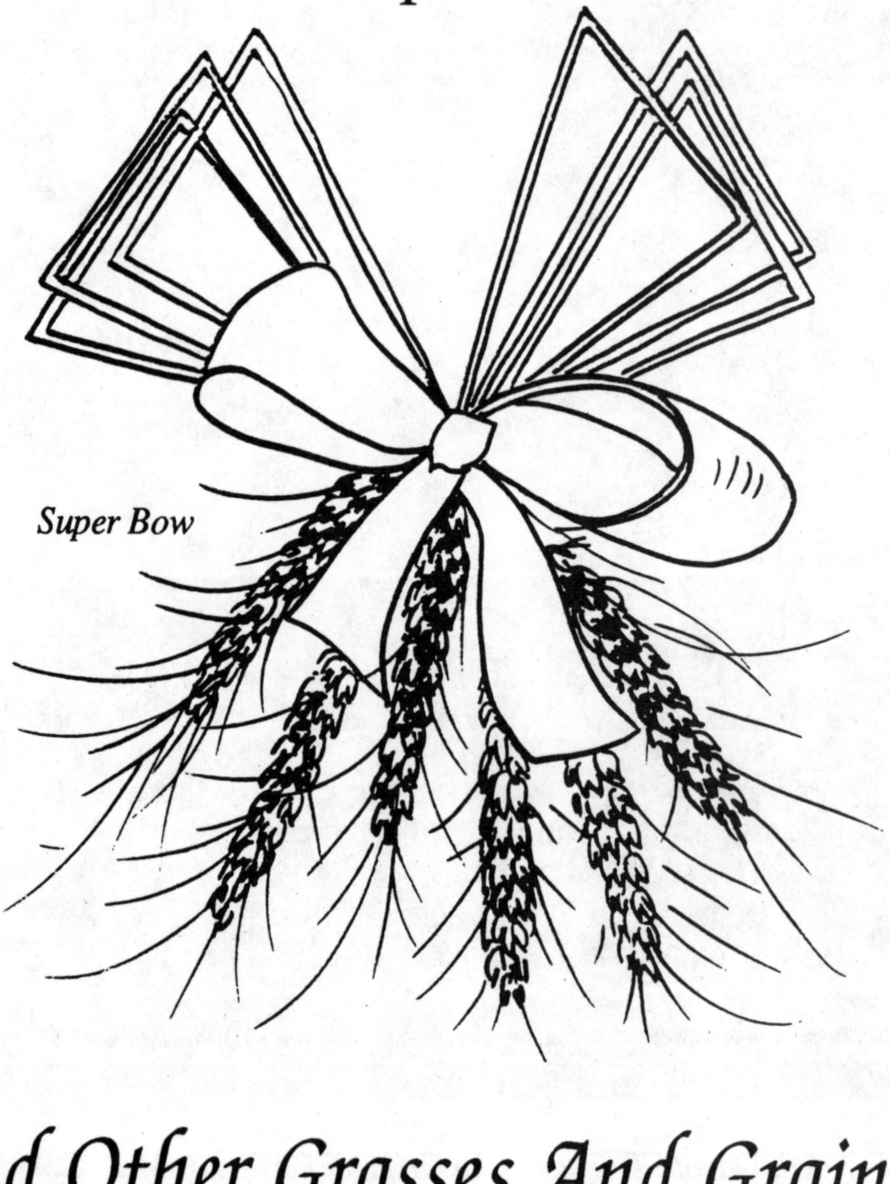

Super Bow

And Other Grasses And Grains

The Ancestry Of Wheat And Other Grasses And Grains

Wild Emmer

Aegilops

Spelt

Wheat has been used as a domestic crop for thousands of years. Today there are many kinds of wheat found within the winter and spring wheat varieties, all of which were derived from a species known as "spelt." Many botanical geneticists believe that "Wild Emmer" and "Aegilops" were the original procreators of spelt.

Rye and Oats were developed as domestic crops from weeds which grew in the wheat fields. These two crops thrive in a fairly poor soil, one that is not rich enough to produce more grain per acre at a lower price than any other grain.

Description Of The Grains

WHEAT is the most flexible of all the grains and for this reason is the standard choice when making Wheat Dollies.

BEARDED WHEAT is distinguished by having wheat heads with four spikelets containing fine, long beards.

UNBEARDED WHEAT usually looks shorter, as it has a more stubby, sturdy distinctiveness about it.

OATS have a lovely shine which attracts added attention. This along with its graceful, un-uniform grains is excellent for finishing touches to the Wheat Dolly.

Ancestry Of Oats And Rye

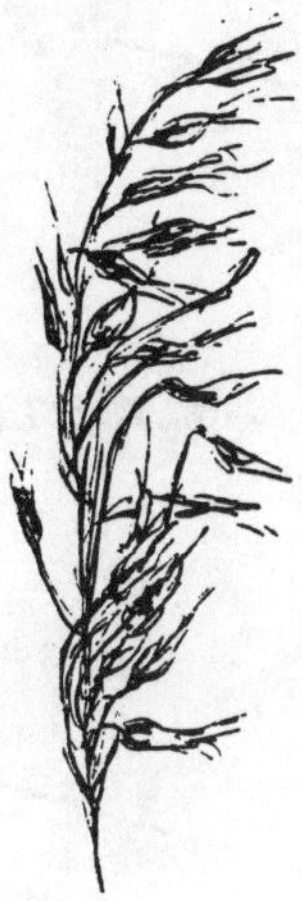

Oats

Rye

Wild Oats

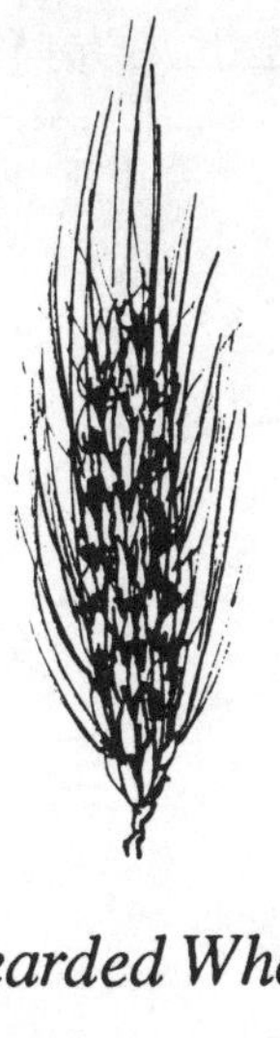

Bearded Wheat

Unbearded Wheat

Barley

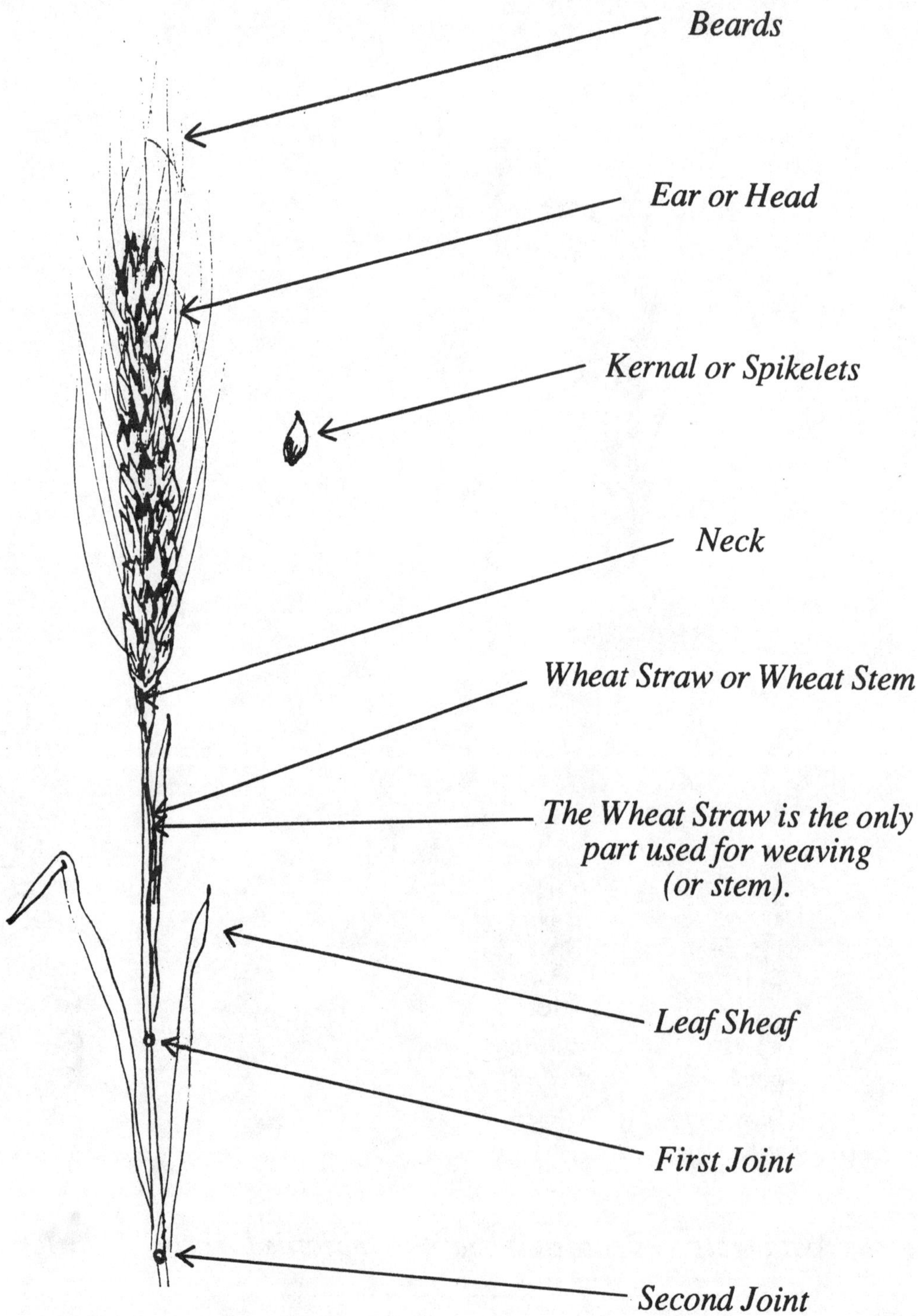
Beards
Ear or Head
Kernal or Spikelets
Neck
Wheat Straw or Wheat Stem
The Wheat Straw is the only part used for weaving (or stem).
Leaf Sheaf
First Joint
Second Joint

Collecting Wheat And Grains

There are several ways to obtain the wheat you need:

(1) Farmers are usually happy to sell or give you small amounts.
(2) It can be found along the roadside.
(3) Bought in floral supply shops.
(4) Grown in your garden in as small a space as 4 x 4 feet.

When gathering wheat, choose a warm, sunny day. Look for tall, upright, pliable stems which will make joining easy. Cut stems as close to the ground as possible.

Sorting And Storing Wheat

It is advisable to wear gloves to protect your hands from "wheat splinters." Clean each stem by removing the sheath, and sort it according to length. Use quart jars and stand the stems upright as this makes sorting easy.

If you plan to store the wheat and use it later, put it into small bunches of approximately 1/2 to 1 pound. Tie the bunches at the bottom, and hang them on a line to dry thoroughly before storing.

Preparing Wheat

Wheat gathered immediately from the field, needs no soaking before weaving. However, wheat that has been dried becomes brittle and will easily split unless properly soaked. Previously dried wheat must be submerged (stem and head) in water for about 30 minutes, until it is flexible enough for weaving. Place any heavy object on top to keep it from floating (a brick works very well). When the wheat is removed from the water, wrap in a damp towel. It can be kept in this manner for as long as twenty-four hours. Wheat stems should not be left in water to become oversoaked, and they should not be re-soaked more than twice.

Basic Instructions For Growing Your Own Wheat

CHAPTER 2

Petit Wheat Angels

Basic Instructions For Growing Wheat

There is something special about making Wheat Dollies with the wheat that has been planted, cared for, and harvested yourself. An eight by eight foot area will yield enough wheat to last an entire season for the beginningWheat Dolly designer. It is very easy to grow.

Wheat grows well in almost any soil and almost anywhere in the United States, provided it receives plenty of water and has about two to three months of warm weather. All you do is prepare average good soil in a sunny area, and sow seeds directly into the ground where it is to grow. The seeds should be planted close together. This will help the wheat to be supportive of each other when the heads begin to develop. Cover the seeds with an inch of grass clippings. Keep the ground lightly watered, but not soggy. If nature does not provide enough rain, a good watering once a week will suffice. Water gently, overhead until the wheat is approximately two or three inches high. After that, water by irrigation only.

Fence off the area with four strong posts three feet high, and tie strong string from post to post. This will help to support the wheat as it grows and protect it from strong winds and animals.

It takes approximately two months to grow wheat and another month to dry it. You may harvest the wheat when it is slightly green or it can be allowed to ripen to a golden yellow color before it is cut.

Preparing The Wheat

CHAPTER 3

Rope Knot

To get started you need:

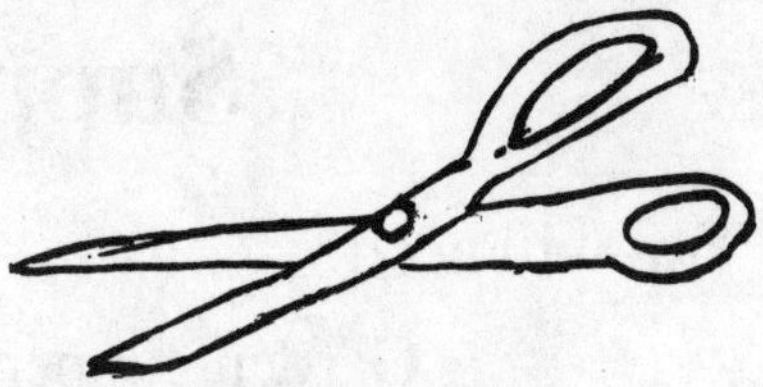

Scissors

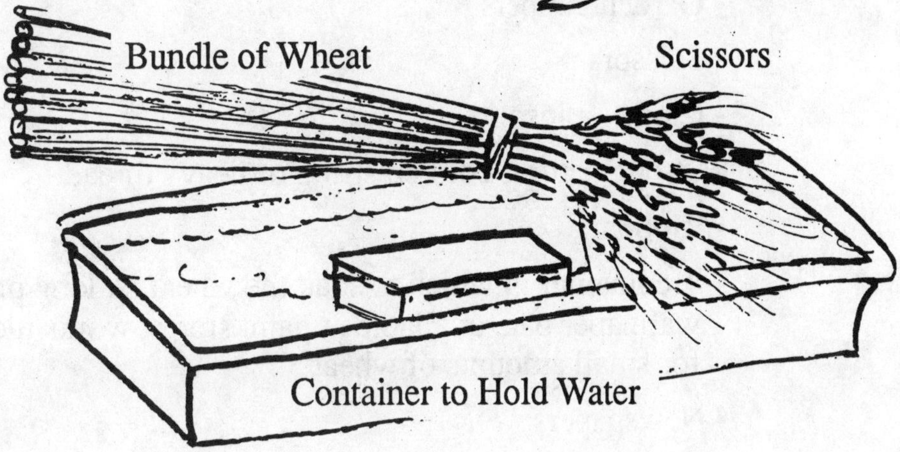

Bundle of Wheat

Container to Hold Water

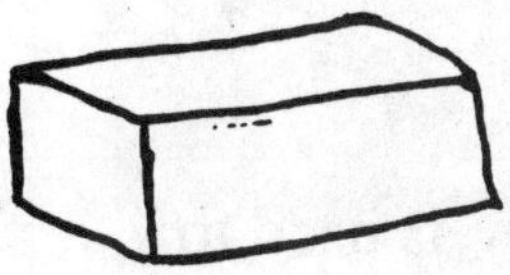

Brick, weighing down wheat underneath water as it soaks.

Towel to wrap wet wheat in after it has soaked.

Sun

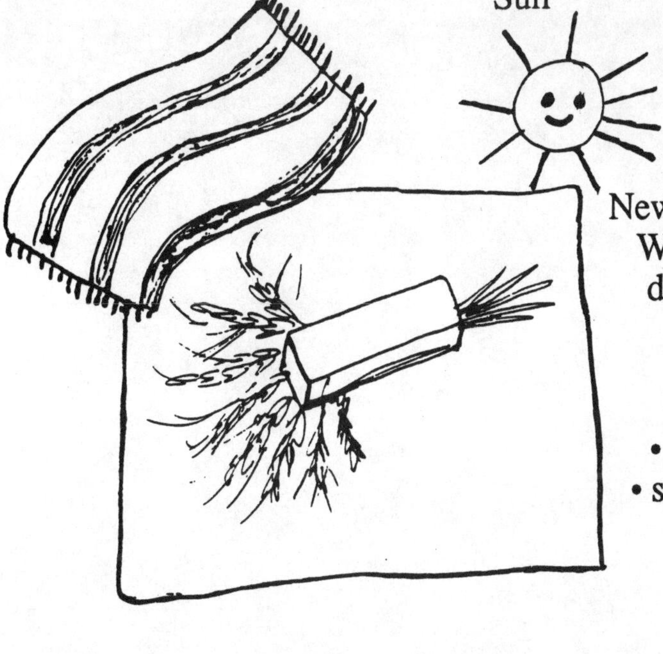

Newspapers underneath and on top of Wheat Dolly. Weigh Wheat Dolly down with a brick to keep it flat. Dry in full sun.

- Strong string – natural color
- wire – several different gauges
- soft cotton rope – 3/4" to 1" thick
- red or green *strong* string
- assorted ribbons
- raffia

Supplies Needed

You will need:

- Garden clippers
- Scissors
- Clothes-pins
- Strong natural-colored string or heavy thread
- Ribbon
- A container in which to soak the wheat. A long plastic wallpaper pan, available at paint stores, works nicely for small amounts of wheat.
- Newspapers
- Several bricks
- Towel
- Wire of various sizes. Heavy (18 gauge) wire is used to create shapes including circles, squares, arms, legs, stars, as well as other designs.
- Tape measure

Care Of Your Wheat Dolly

CHAPTER 4

Framed Wheat Dolly

Wheat Dolly Care

Wheat is strong and durable, demonstrated by the fact that the grain found in ancient pyramids was in remarkably good condition. Today's Wheat Dollies could be enjoyed by future generations with only a little care given them. Treat them as you would any other ornament in your home. An occasional dusting with a small brush is all that is necessary. For a more thorough job, rinse well with a solution of warm water and Borax or Clorox. Sun or oven (120 degrees) dry immediately. Adding a fresh ribbon where one is needed will keep your Wheat Dolly looking fresh and new.

If you wish to store Wheat Dollies for more than a few months, wrap in tissue paper, place in a cardboard box, to which a few mothballs or an open can of tobacco has been added. Place the box in a large plastic storage bag, label, and store in a dry cool temperature, free of rodents. This will keep your wheat in excellent condition until ready to use again.

Basic Weaving Instructions

CHAPTER 5

Two Three-Straw Plaits

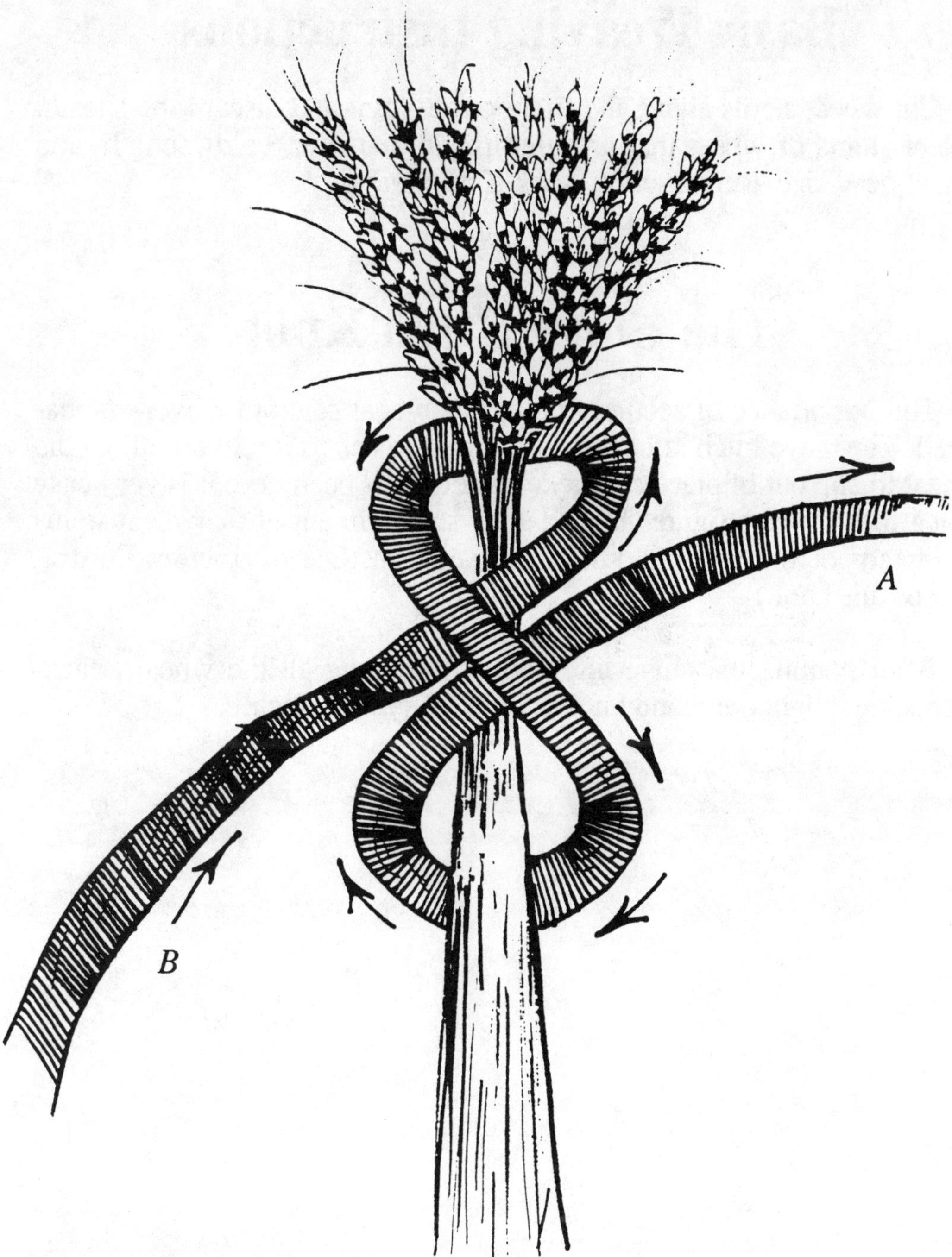

The Clove Hitch Knot

Basic Weaving Instructions

Cut wheat stems above the first bottom joint and discard the sheath. Select stems of wheat that are uniform in diameter, color, length, and wheat head size. Remember to soak if necessary.

The Clove Hitch Knot

The importance of securely tieing the wheat cannot be over-emphasized. The clove hitch knot is the very best to use as it will not allow the wheat to slip out of place or slacken after it has been tied. It is very easy to learn. It forms a figure "8" and allows adjustment of the wheat stems and heads before the final knot is made tight. (See the picture illustration of this knot.)

After making the clove hitch and completing all the wheat adjustments, tie a tight overhand knot to hold everything secure.

Joining Wheat For Continued Plaiting

Wheat head has been cut off.

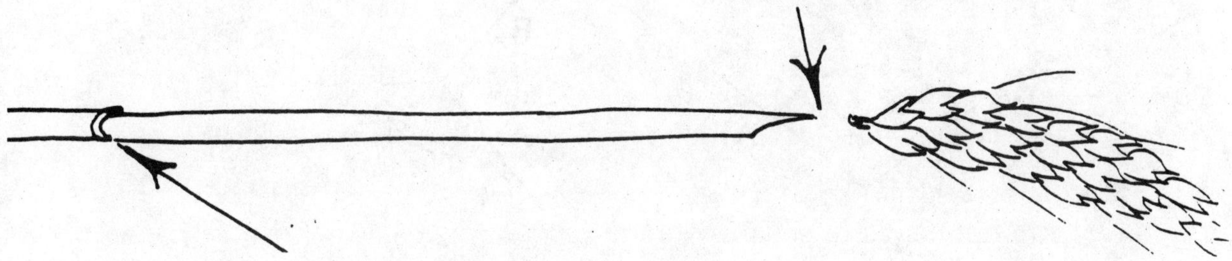

Cut just above the first joint.

Wheat Stem Joining

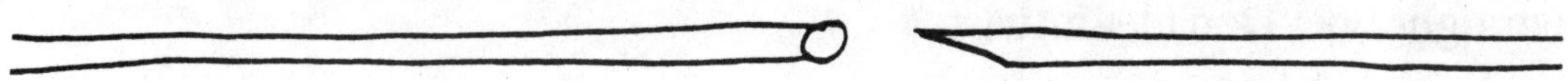

You are now ready to continue to weave.

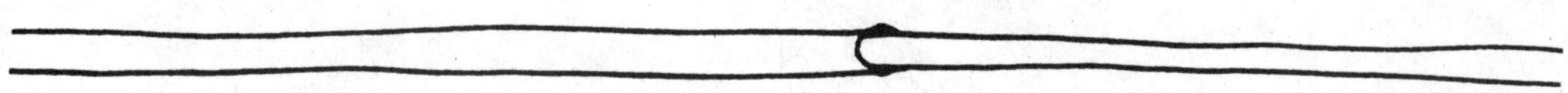

Only the upper part of the wheat stem is flexible enough to use for plaiting. *Avoid* extending (or joining) all the wheat stems with which you are working in the same area, as it will weaken the plait. Stagger the positions instead.

Two Straw Plait

Tie two wheat straws at neck of wheat and place at 120° angle apart.

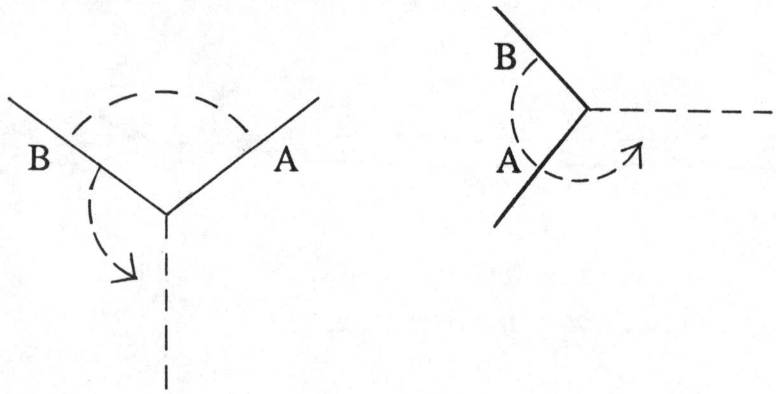

Place the working straw A up and over B and place 120° from B straw. B straw now becomes the working straw. Place B straw (fold) up and over A. Repeat alternating wheat straw each time. Keep plait in a straight line. Don't let it twist.

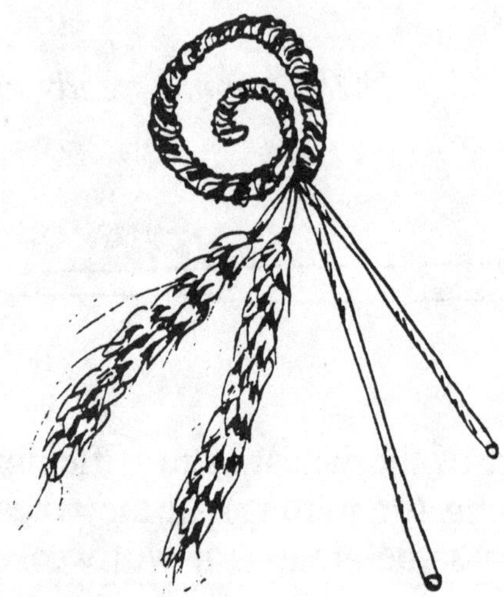

Two Straw Plait

Three Straw Plait

Tie with the clove hitch knot. Secure neck of wheat heads tightly together. This is the simple Three Straw Plait (Three Straw Braid).

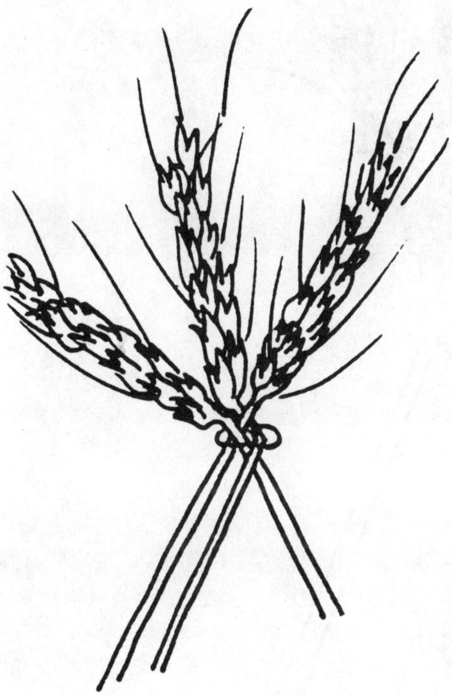

Three Straw Plait

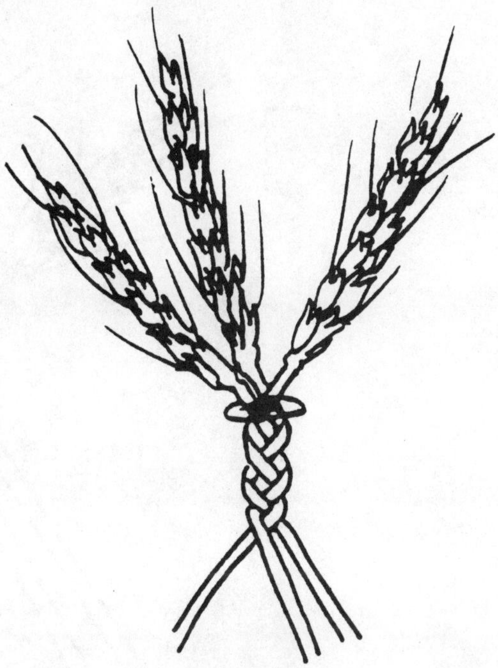

Detail of Common Braid

Detail Of How To Start A Three Straw Plait

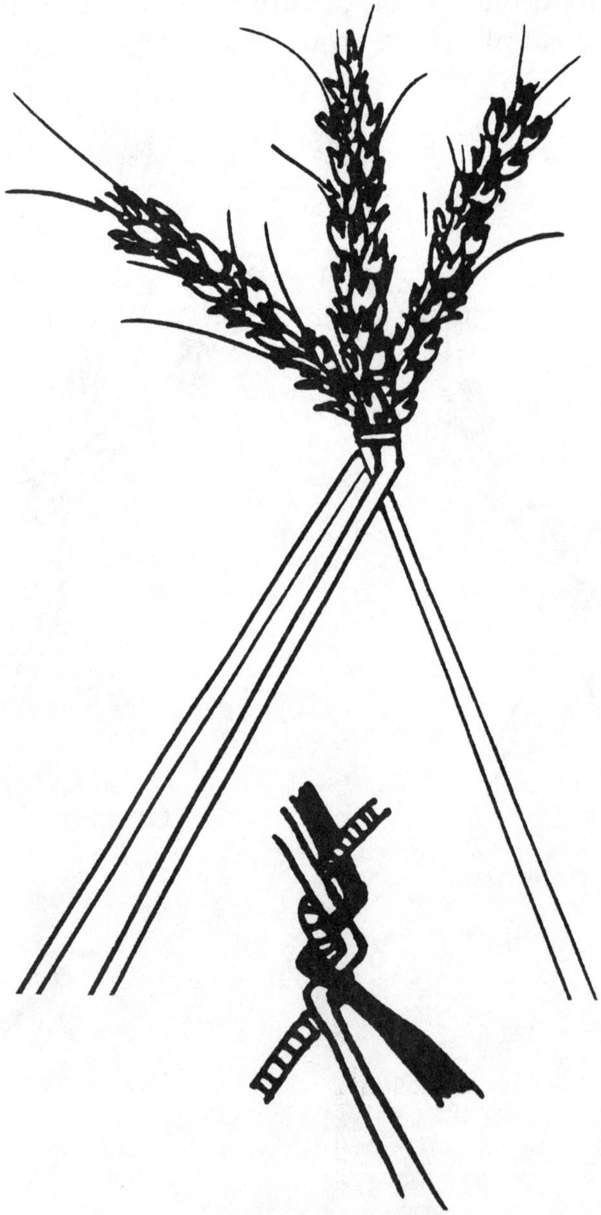

Common Braid

Two Three Straw Plaits

Tie three straws together at neck. Spread stems out in different directions. Straw #1 is placed in #3 position. Straw #2 is placed in #4 position. Repeat this pattern always moving straws #1 and #3 in opposite directions, while #3 straw moves in vacant position.

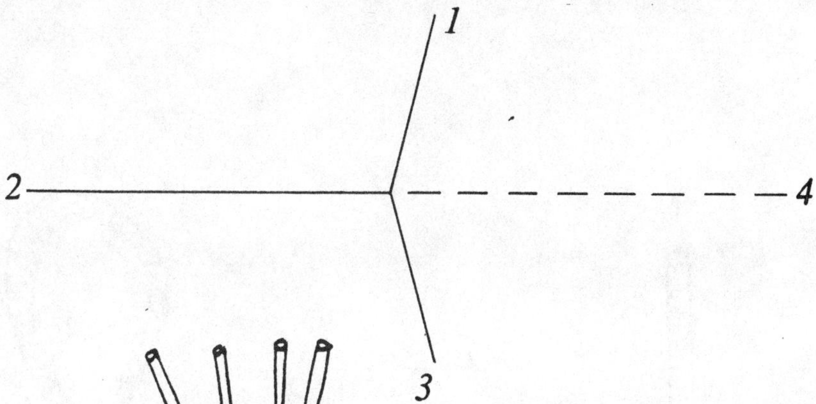

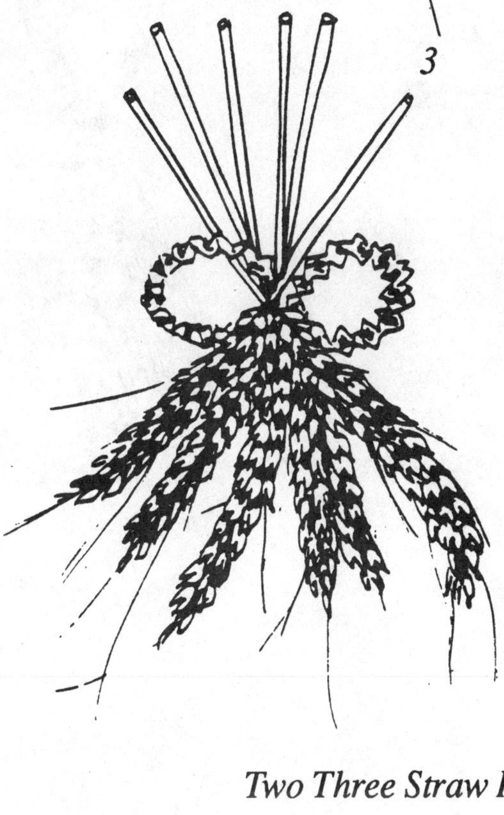

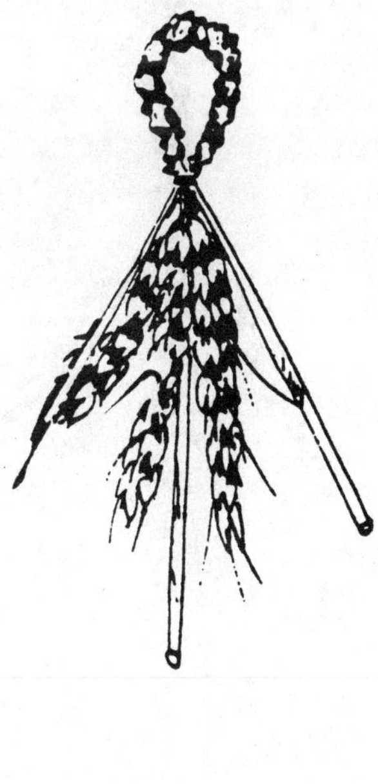

Two Three Straw Plaits

Three Straw Cat's Paw Plait

Tie three stems together at necks. Spread out in N-E-W positions. Place E stem in N position to the right side of the N stem. Then place W straw to N position to the left of the N stem. Continue this same pattern until desired length has been obtained.

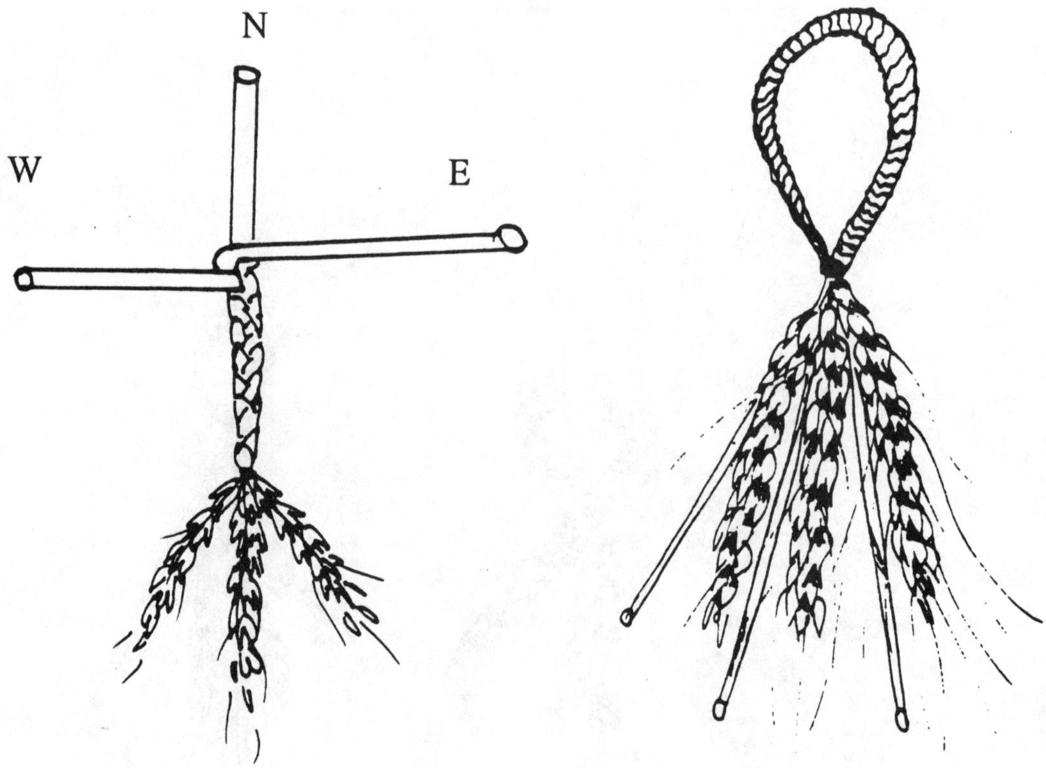

Three Straw Cat's Paw Plait

Four Straw Plait

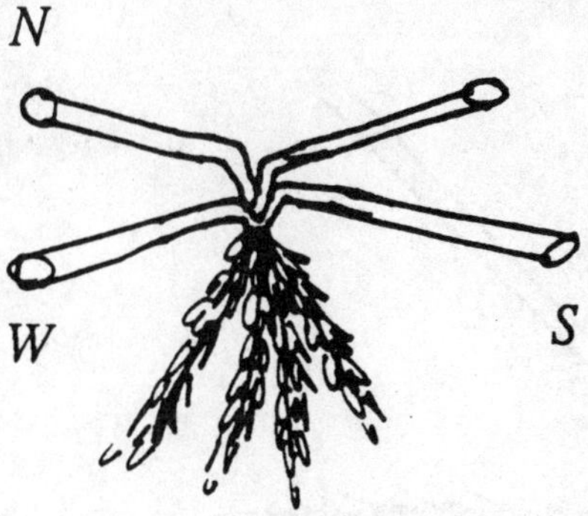

Tie four straws of equal length at necks with heads facing down. Spread straws.

Hold thumb on top of straws to keep in a tight position – as you place N straw to S and place S straw up to N. Hold firmly with thumb up to N. Firmly hold and move E straw to W and W straw to E. Press firmly. Repeat moving in the same order.

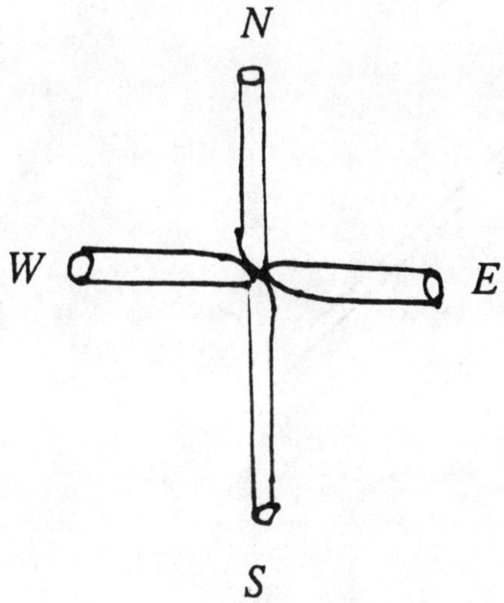

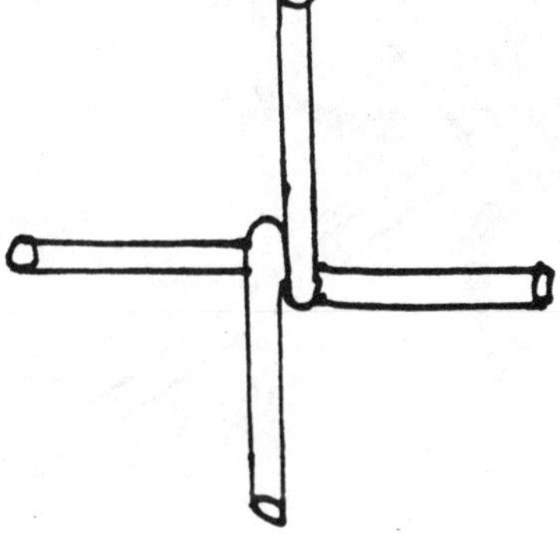

Four Straw Plait

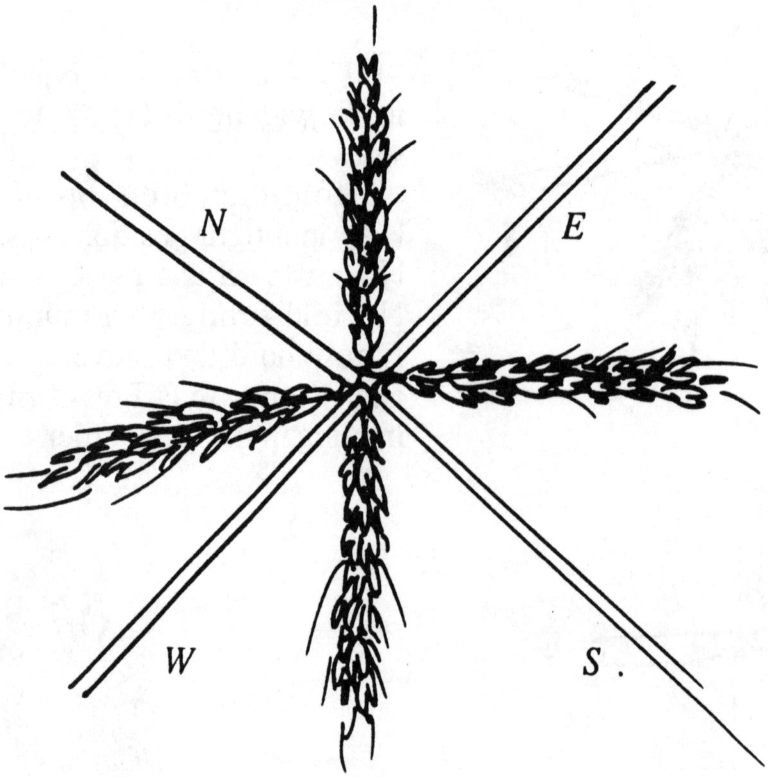
N
E
W
S

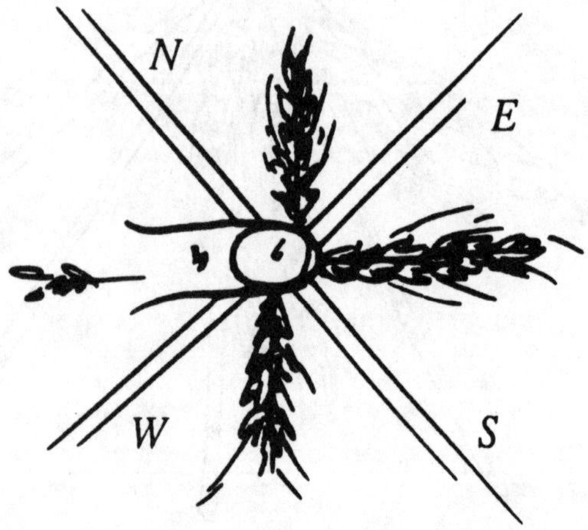
N
E
W
S

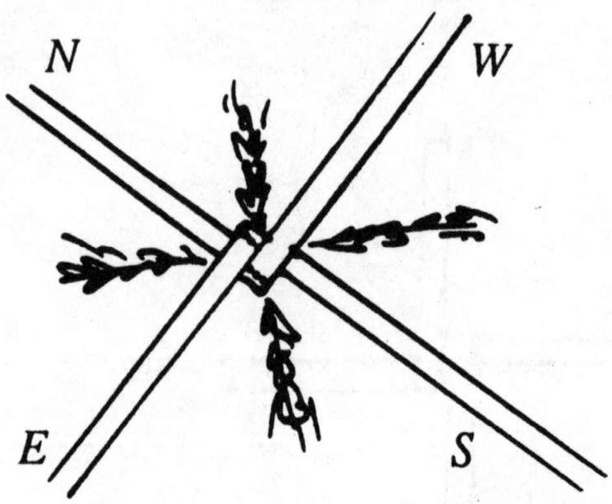
N
W
E
S

Five Straw Plait

Tie five wheat heads at necks. Lay flat on table top or counter three straws to right and two straws left. Place straw #1 under #2 and over #3 to lie next to straw #4. Place straw #5 under #4 and over straw #1 to lie next to #3. Continue pattern taking outer straw under and over next two straws. Working right to left sides alternately.

Step B

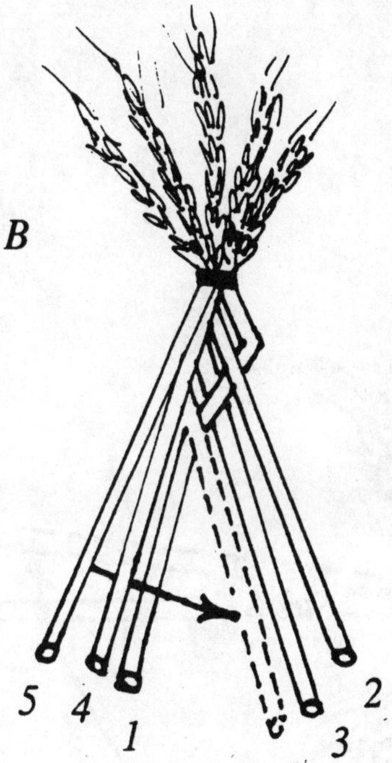

Step A

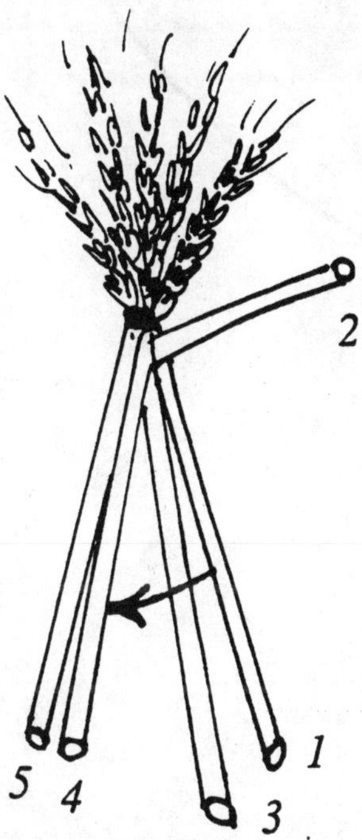

Five Straw Plait

Five Straw Plait

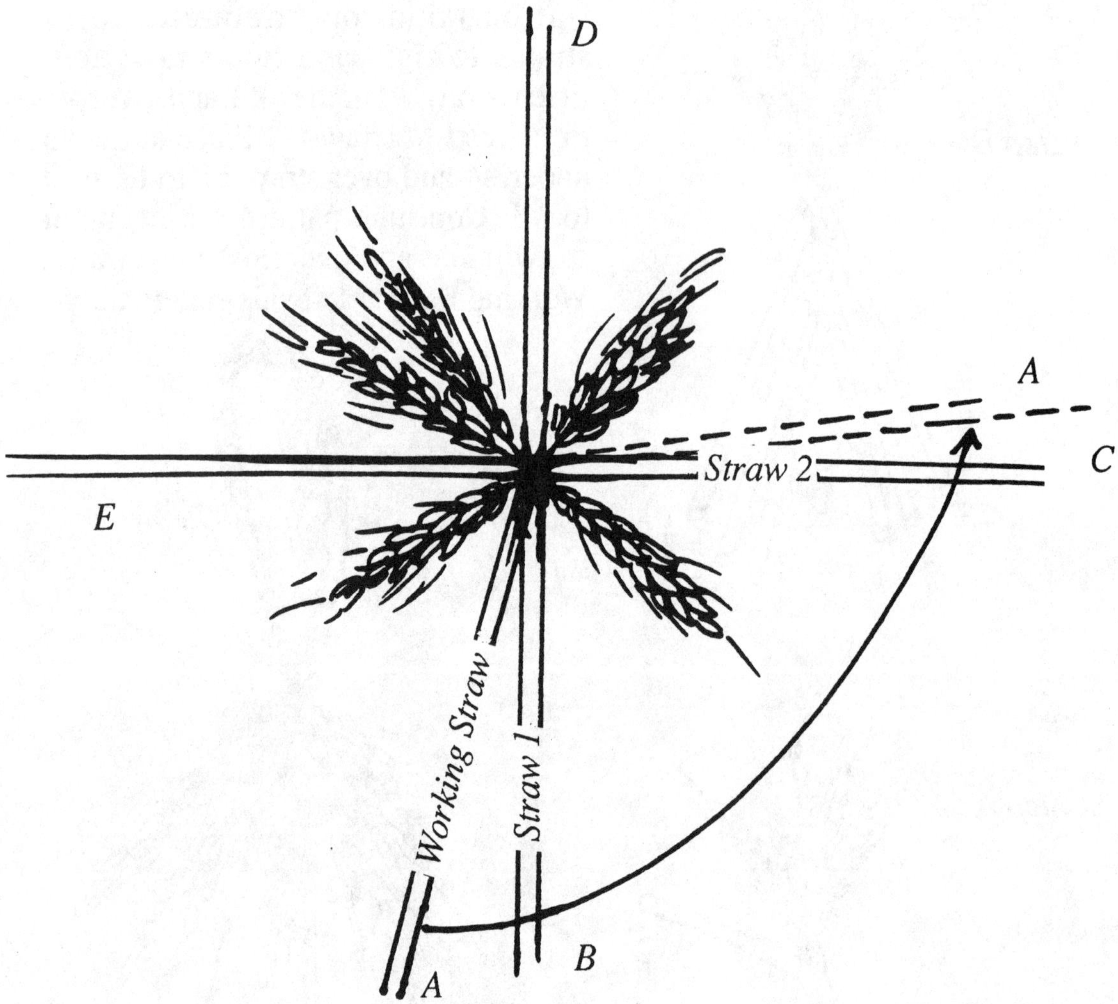

Six Straw Round Plait

Tie six straws together at necks spread out in six directions. Hold flat with thumb and forefinger. Beginning with any straw, place over two straws in a clockwise direction. (The beginning straw is called the working straw.) As you rotate work, you will have a "new" working straw to place over two more straws. Continue this pattern until desired length has been reached.

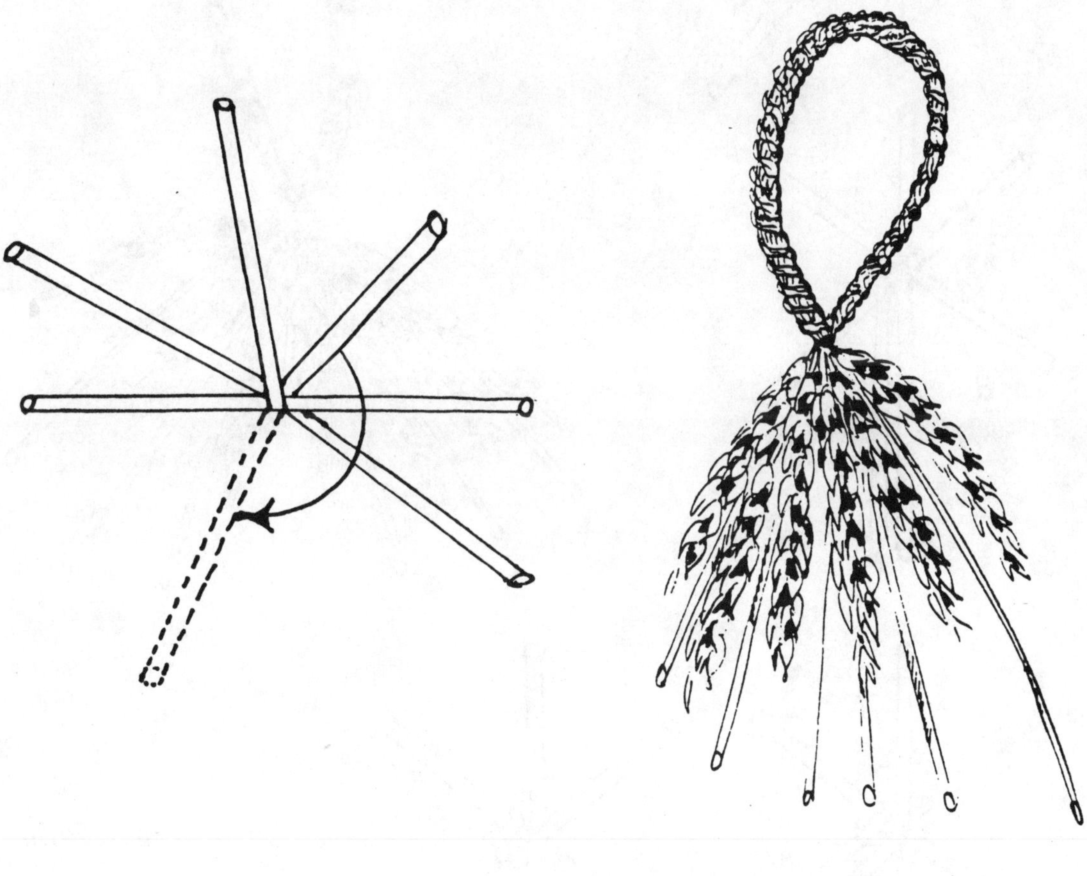

Six Straw Round Plait

Seven Straw Round Plait

Tie seven straws together. Plait as you would for a five straw plait, but be sure to have six corners. Rotate one-sixth of a circle. Six folds in a circle.

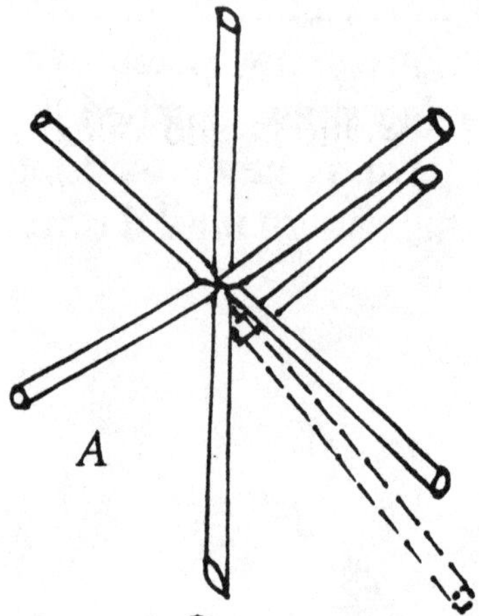

A

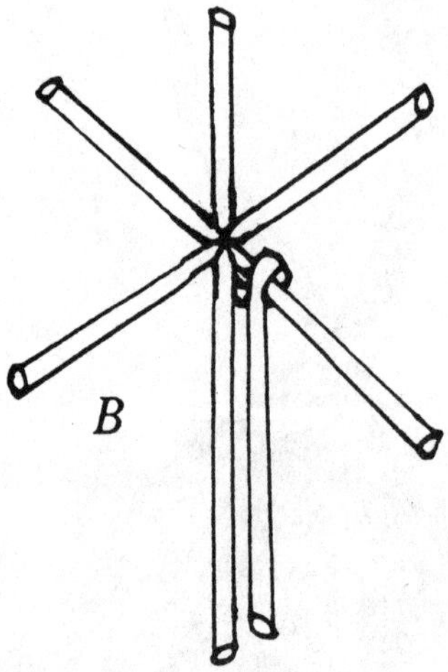

B

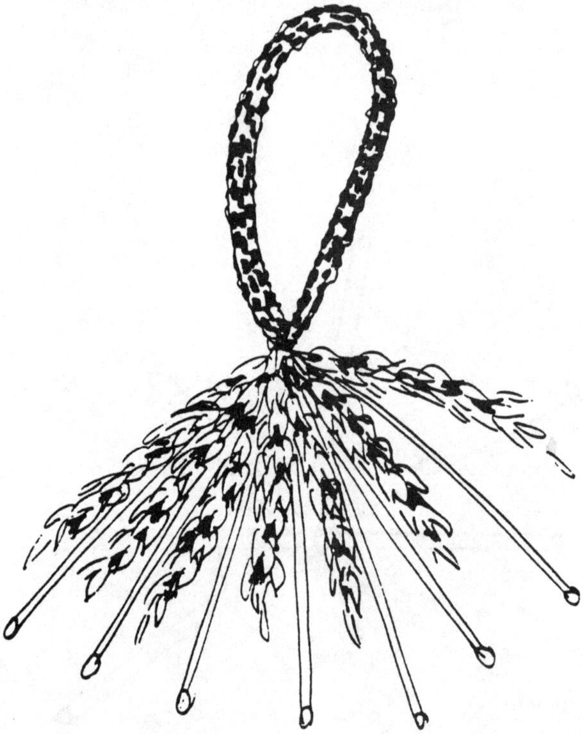

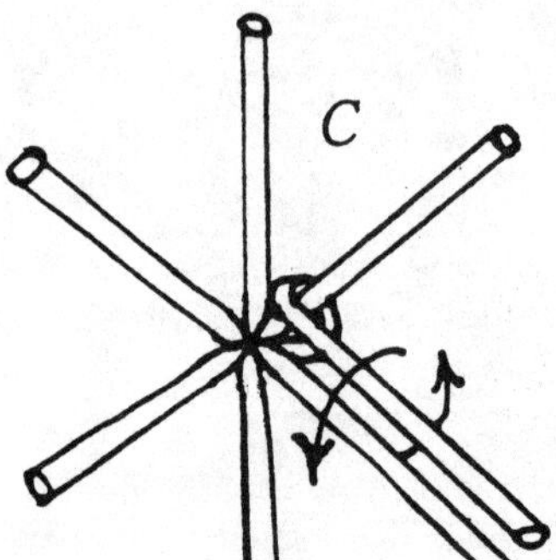

C

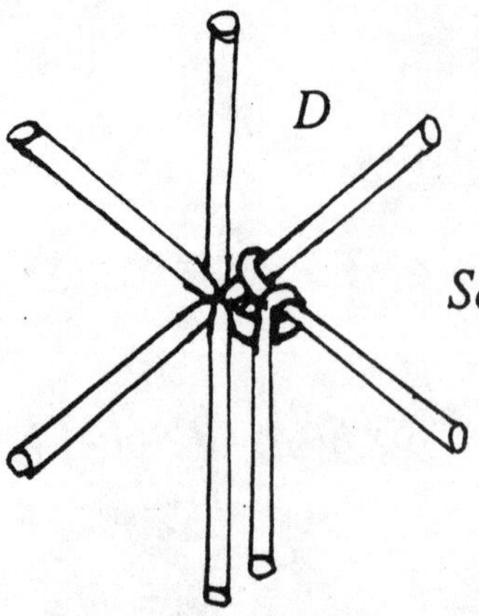

D

Seven Straw Round Plait

The Seven Straw Round Plait

A chain like pattern. Tie seven wheat stems together at the neck of the heads. Place straw #1 in front of straw #7, around it and place it next to straw #3. Next place straw #4 behind #5, #6, and #7 stems – around front of #7 and place it next to #6 straw. Next place #2 straw in front of straw #7, around it and place it next to #1 straw. Take #5 straw behind #7, then across the front and place next to #4 straw. Continue with this pattern to the desired length.

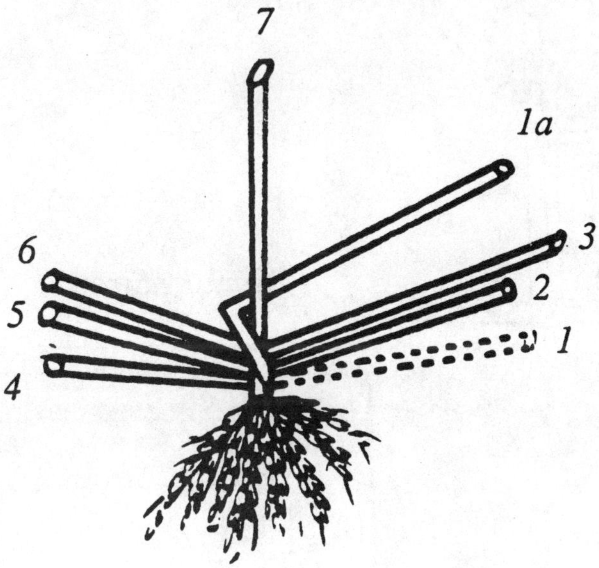

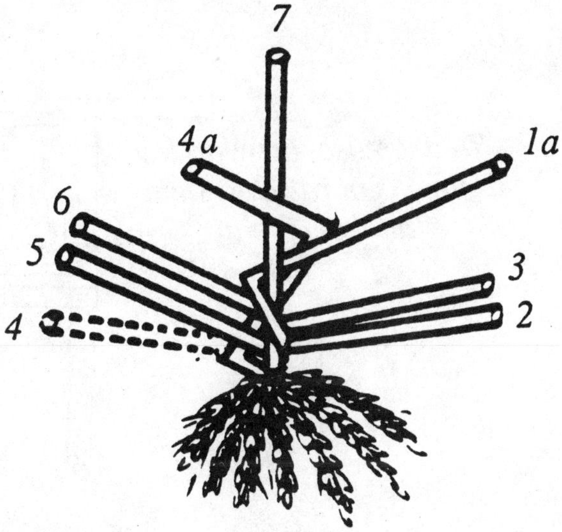

Increasing And Decreasing

When working straw goes on top of parallel straw, the work stays the same size.

To increase, the working wheat stem is to right of parallel stem.

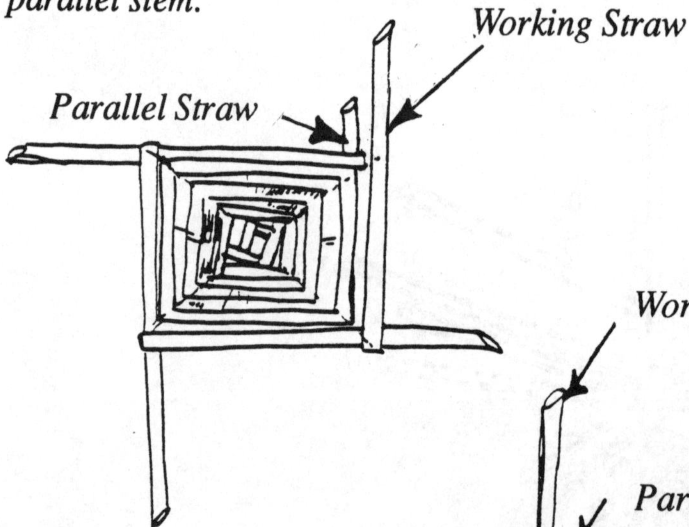

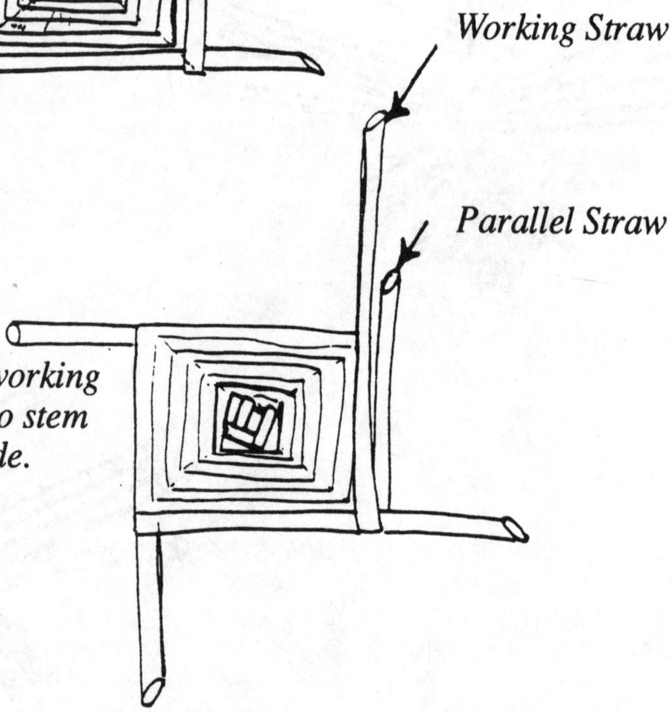

To decrease, the working stem is parallel to stem on the left side.

Designing With Plaits

The beginner will have a wonderful time inventing all sorts of unusual ways to put the newly discovered plaiting into effect, such as with crowns, straw ladies hats, baskets, circles, halos, arms, heads, picture frames, mirror frames, hearts, many in unusual curves and bends, or instertwined, knotted, braided, twisted, circulared are only a very few I'll mention. The list is as long as one's imagination. There are no set rules you must follow.

You might want to try the **Olympic Circles** – five circles intertwist and attach in place with glue or tiny thin wire.

Plaited Circles – fasten ends together to form a circle, add ribbon and wheat stems and wheat heads.

Simple Straw Maiden – The arms are made with a four or seven plait. Twenty-five straws in a downward position become the body. Insert upper arms through half of the body straws and tie. Attach hands and heads at the front. Plait halo and tie.

Olympic Circles

Plaited Circles

Simple Straw Maiden

Hearts Intertwined

Hearts Intertwined is done with a thin flexible wire inserted. When the desired length has been reached, shape into the form of a heart. Hearts can be attached (intertwined) to each other before the ends of the hearts are tied together with raffia. Bows can very often be placed anywhere that the ends meet.

Hearts Intertwined

Parallel Plait

To keep the plait an exact width, the working straw is placed directly on top of the parallel straw and the work becomes longer, but not wider. This is used for making handles on baskets, arms for figures, circles, halos, stars, ropes, etc.

Increase Plait

This plait will become wider and larger in size, when the working straw is placed to the right of the parallel straw. Increase plaiting is needed when making the Cornucopia, the Bell, the Spiral. These are just a few examples of its use.

Decrease Plait

Sometimes the plait must be made smaller once it has been increased. The working straw is placed to the left of the parallel straw. The Wheat Dolly figures are good examples of the use of the Decreasing plait after it has been increased.

Reverse Plait

Begin by increasing the plait to the desired size. Stop increasing the plait and lay the wheat on top of the second stem. Bend the wheat stem down and back to the position it came from on the reverse side. Turn the wheat dolly over with the work facing downward. Bend the wheat stem over two and repeat 3 more times. The wheat dolly will be in position to continue with the four straw plaiting. The Lantern is an example where the Reverse plait is used.

Flat Ribbon Plait

This ribbon plait is often used as a trim on baskets or picture frames, as a book mark, or on wheat figures.

1. Tie nine straws together with the clove hitch knot.
2. Divide straws so that four go to the left and five to the right.
3. Lift the second straw on the right side.
4. Bend over the first straw on the right so it becomes parallel to the straw on the left (inside of the opposite side).
5. Bend second straw parallel to the right side. Repeat, locking both sides as you plait.
6. Flatten the wheat straw with the dull side of a dinner knife.

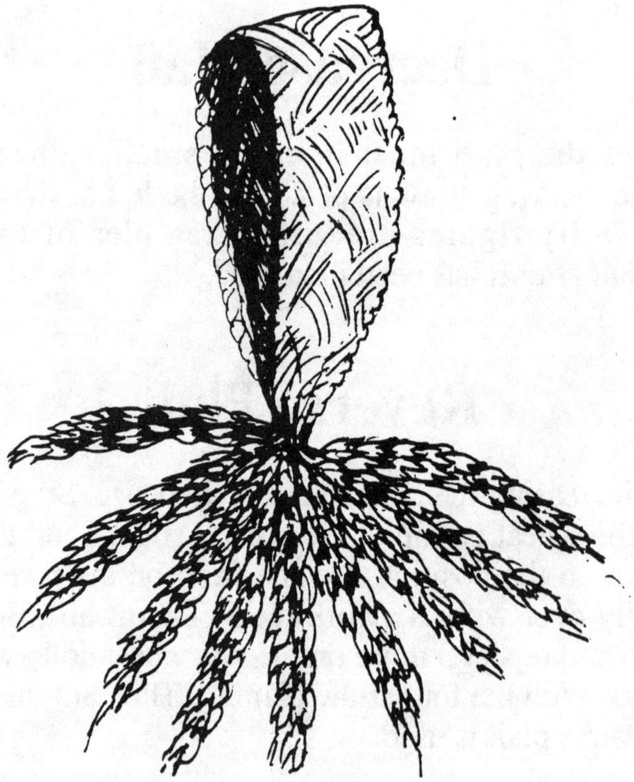

Plaiting With A Core

The core is made of different materials, depending on what you plan to make. Some wheat plaiters use only wheat for a core, while others add rope and wood dowels to that list. A rope is easy to shape into an oval, round, or half circle. Wood is used for candles, crosses, and straight woven wheat dollies.

1\.

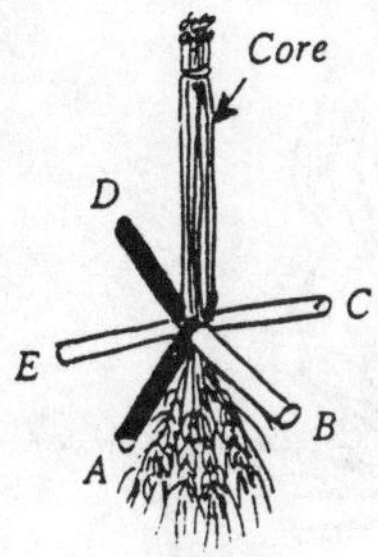

Tie a bunch of straw together at the neck of the wheat heads. Use the clove hitch knot. Select five straws that are uniform in size spread them out flat in front of the worker, the wheat heads pointed downwards. A and B lie directly in front of worker.

2\.

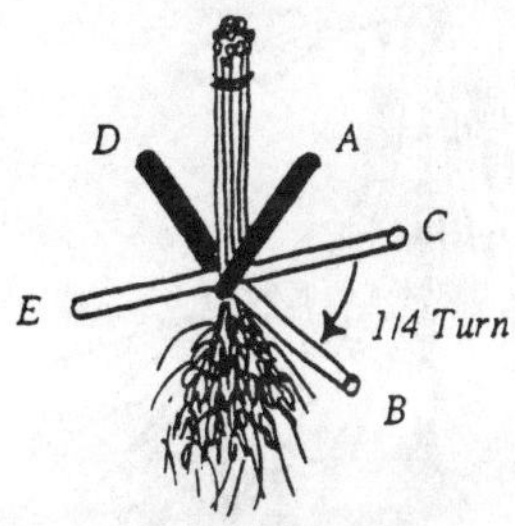

With your left hand, hold the core with your thumb and first finger. Move A over B and C with right hand using thumb and first finger. With a quarter turn clockwise, A and B will lie directly in front of worker. Work close to point of core.

3\.

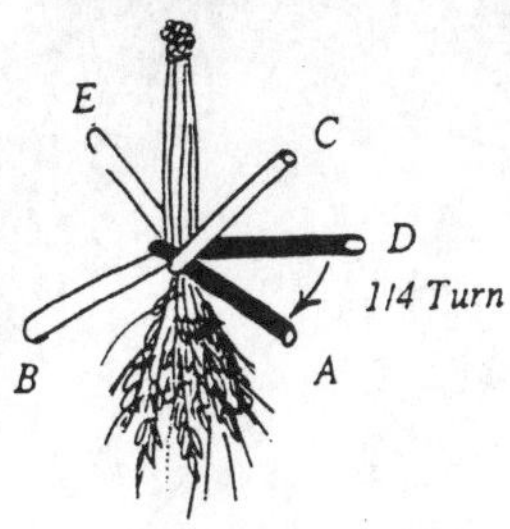

Move C over A and D, and make another quarter turn clockwise. As you repeat these movements, with straws D and E a square or box is formed around the center core.

4\.

Note the square or box being formed around the core of wheat. The core can be made out of a rope, wire and rope combination, material, or it can be a hollow core. Experimentation will help you decide which core is desired.

Beginning Wheat Dollies

CHAPTER 6

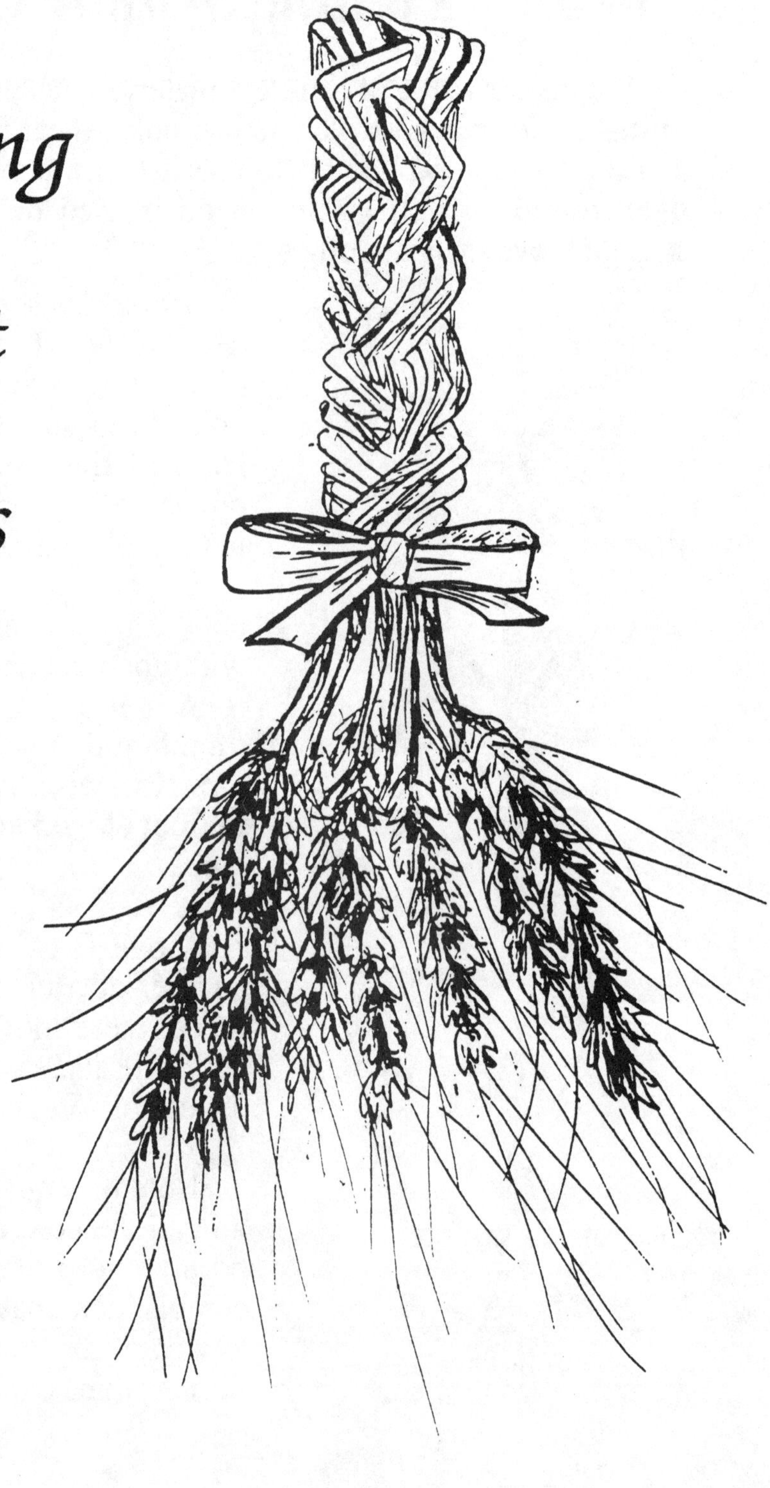

Gem

North Dakota Treasure

North Dakota Treasure

The bountiful North Dakota Treasure is one of the most successful Wheat Dollies the beginner can make, yet its appearance is eye catching.

1. It requires seventy-five wheat straws with large full heads.
2. Soak the wheat for the required time (see page 17).
3. Divide the wheat into three bundles of twenty-five straws each. The center upright wheat straws should be longer in proportion to the cross member wheat straws.
4. The cross-section is assembled with wheat heads at both ends. The wheat stem ends are then tucked behind the heads at the backside and then tied securely with the clove hitch knot.
5. The center upright section of the wheat straws is tied just below the heads at the neck with the clove hitch knot.
6. Half of the cross section wheat stems are pulled apart. The upper half is tied at the neck – of the center – upright wheat stems.
7. The lower half of the cross-section is pulled apart about two inches down and tied securely to the upright section.
8. Place on newspapers. Flatten and fan out all of the wheat heads. Cover with more layers of newspapers weighted down with bricks. Dry in full sun for several hours.
9. When completely dry, decorate with a colorful ribbon.

Straw Cross

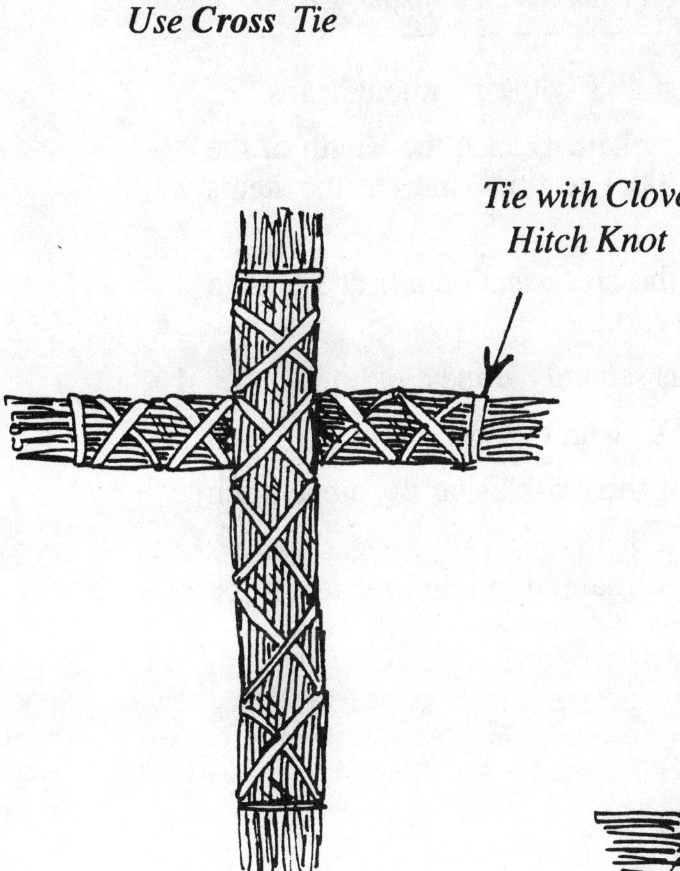
Use Cross Tie
Tie with Clove Hitch Knot

Wrap Wheat Core with String, Ribbon, Raffia, or Wheat Stem.

Straw Cross

The cross has sacred significance at Easter time.
The Straw Cross makes an unusual gift.

1. Select fifty wheat straws with or without heads.
2. Divide the straws in half. Decide the length of the horizontal and vertical sections and cut the stems accordingly.
3. Evenly distribute the cross section wheat stems in with the upright section.
4. Tie the wheat stems securely at the cross mid-section. Form an "X" with the string.
5. Tie all four ends of the cross using the clove hitch knot.
6. A colorful ribbon or natural colored raffia can be used to wrap or cross-tie the wheat dolly cross.

Super Bow

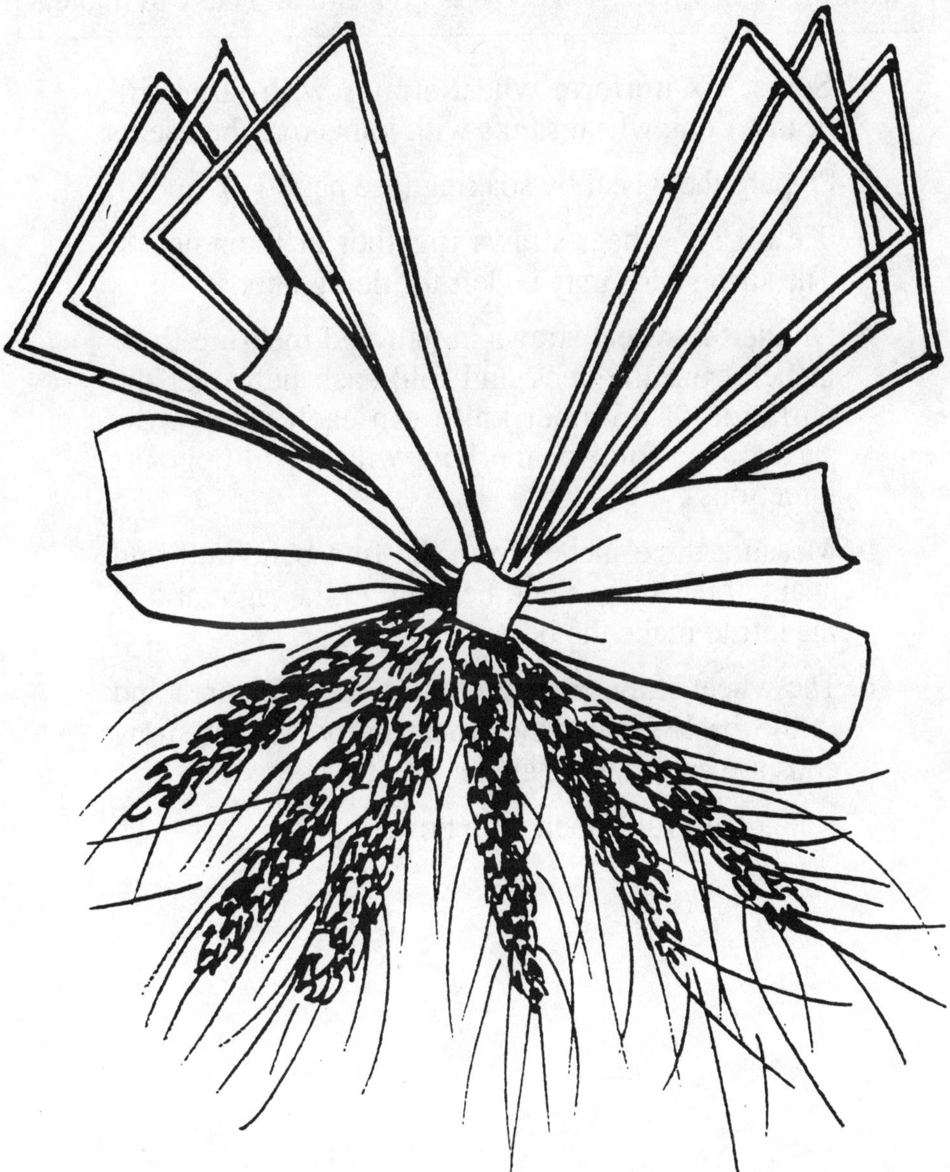

Super Bow

The Super Bow Wheat Dolly is a nice introduction to working with wheat. This bow is both fun and easy to make.

1. Select six uniform wheat straws with large full heads. Long wheat stems with joints may be used.
2. Prepare the wheat by soaking (see page 17).
3. Tie all six wheat straws together at their necks. The string ends may be left for tieing later.
4. Divide the wheat straws in half and measure three inches from the neck and fold each half over the dull side of a dinner knife (on each side of the slant of the fold for the bow will fall in opposite directions).
5. Measure three inches from the first bend and fold again at the same slant, bending to the right and to the left to make the bow.
6. The wheat straw ends are tied to the neck area and to the back side of the wheat dolly. Wheat straw ends are cut off at an angle.
7. Add a colorful ribbon with bow.

Petit Wheat Sheaf

Petit Wheat Sheaf

1. Select fifteen to twenty wheat stems with full heads, and soak the wheat (see page 17).
2. Tie at the neck of the wheat stems using the clove hitch knot. Raffia is strong and attractive and holds a clove hitch knot well. Adjust the wheat heads uniformly.
3. Make an over-hand knot to secure the stems.
4. Fan out the wheat heads and wheat stems. Place the Petit Wheat Sheaf between newspaper and weight it down with bricks. Dry in full sun.
5. When completely dry, add a colorful ribbon of your choice.

Joyful Star #1

Use two bundles of 12 wheat stems. Cut all 8 1/2" long. Soak stems. Tie one bundle together in the center. Use the clove hitch knot and pull very tight so stems will fan out. Now tie the other bundle in the same manner. Tie both bundles together center to center. Fan out uniformly, place between two sheets of newspapers weighted by a heavy object such as a brick and allow wheat to dry. Trim stems at an angle. Ribbon may or may not be added.

These make delightful tree decorations at Christmas time. No other hanging tie is necessary, as they stand on their own on a tree branch.

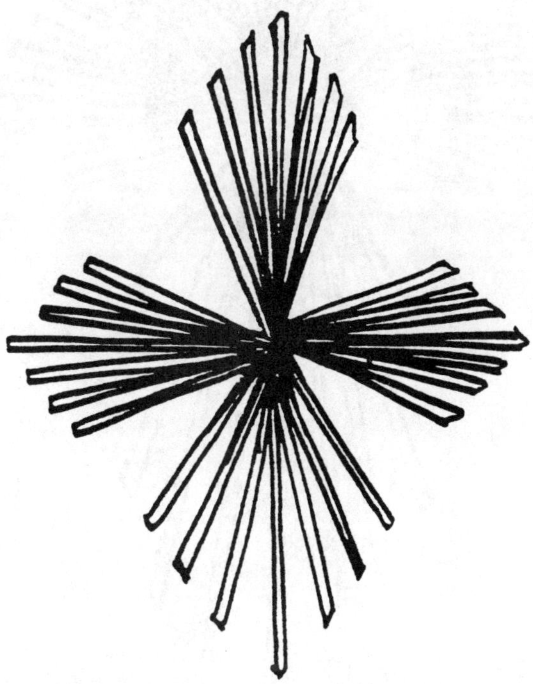

Joyful Star #2

Use directions for Joyful Star #1, using two bundles of 12 stems each. Cut one bundle of stems 8 1/2" long and the second bundle 10 1/2" long. Tie the first bundle in the center with the clove hitch knot. Tie the second bundle 3 1/2" down from one end. Proceed as in Joyful Star #1.

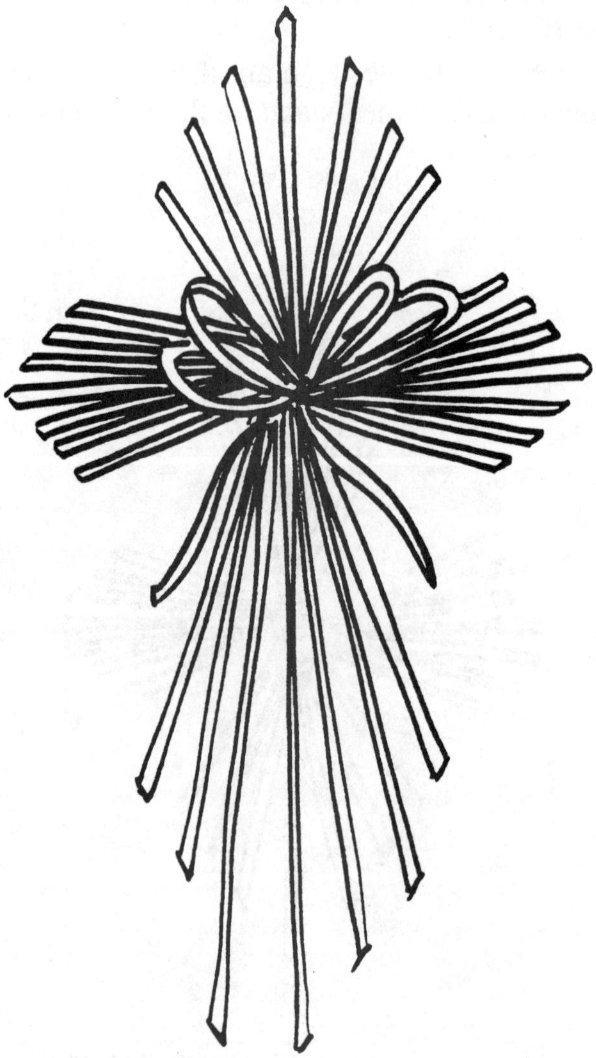

Five Pointed Star

Christmas is an especially nice time to make the Five Pointed Star. Red string adds to the seasonal look.

For different look, use wheat stems with heads attached. Tie the straw ends together instead of the necks.

1. Select 30 wheat stems. Soak the wheat (see page 17).
2. Tie (use a clove hitch knot) all the stems together at the bundle's smallest diameter, nearest the neck. Cut the stems to an identical length.
3. On a flat surface, fold out all the wheat stems to form a circle.
4. Separate the wheat straws into five bunches with six stems each.
5. Tie the ends of each bunch (5) securely with the clove hitch knot.
6. Measure one-third of the stem's length from the center tie on all five bundles. Separate each bundle in half (three straws each). Tie the three stems to the neighbors three stems on each side. Continue all the way around. This forms the star.

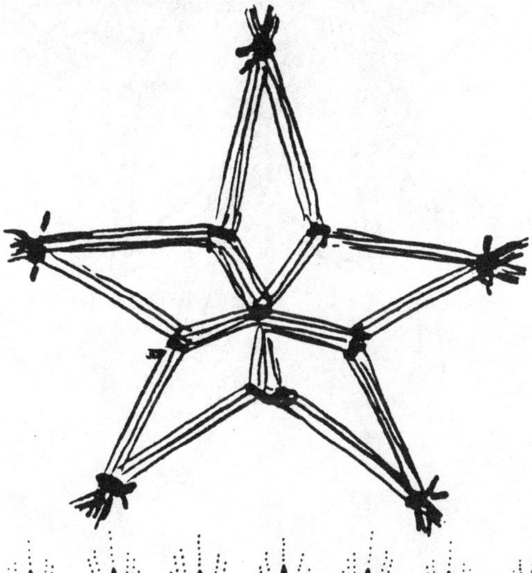

Five Pointed Star With Wheat

Delightful stars to hang on the Christmas tree.

1. Cut ten sturdy wheat straws exactly the same length.
2. Pre-soak the wheat (see page 17).
3. Run a needle and thread through the hollow stem and pull tight enough to let the ends of the wheat stem meet. A drop of glue at each corner will help to secure the wheat stems. (A thin flexible wire will also give support to the star.)

Or:

3. Insert a thin wire into all ten pieces. Bend the wire to form a star.
4. With tiny ribbon and three wheat heads, secure to the star.

Radiant Star

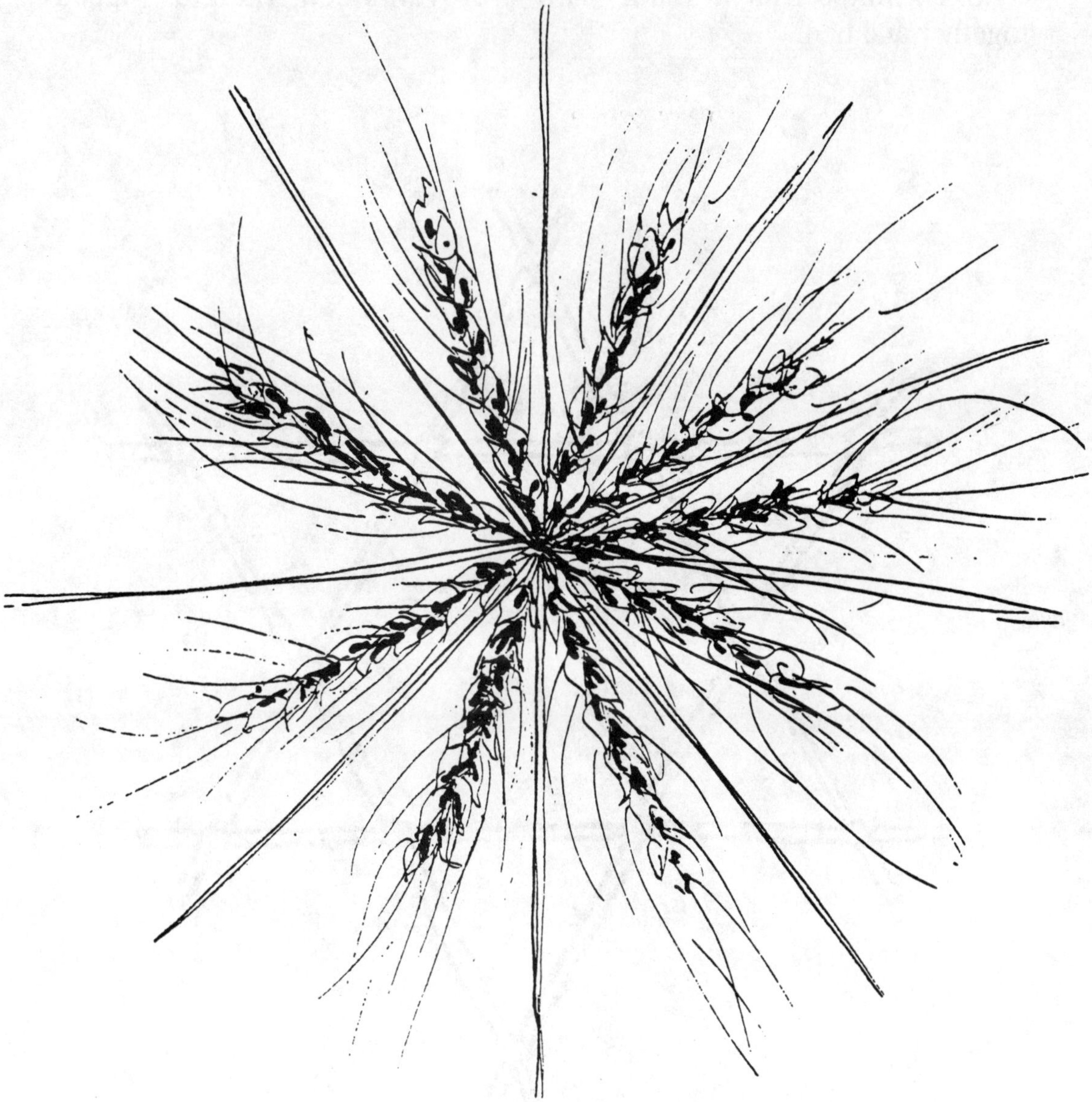

Tie eight wheat stems of uniform length together at the necks, bend back and flatten out. Dry between newspapers and weight with a brick.

Six Pointed Star

Six Pointed Star is made with two even-sided triangles placed together and tied.

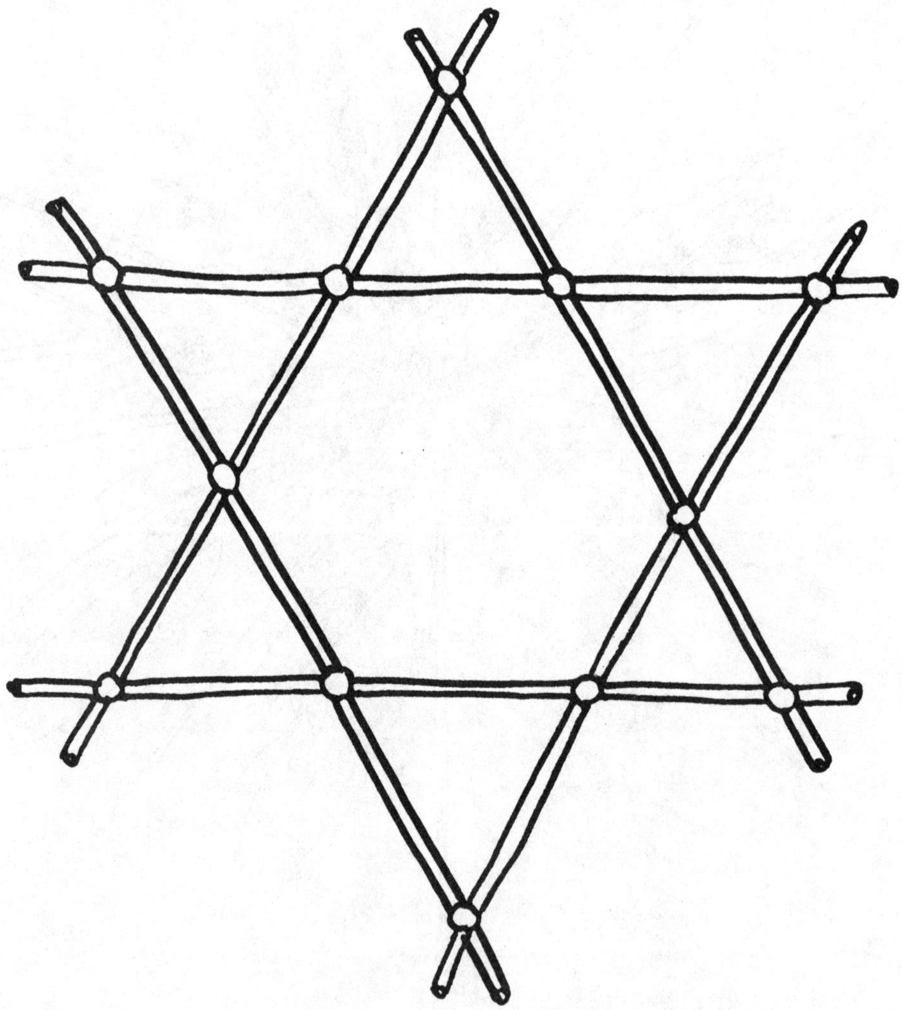

Eight Pointed Star

Sixteen Straws interlaced and tied at ends.

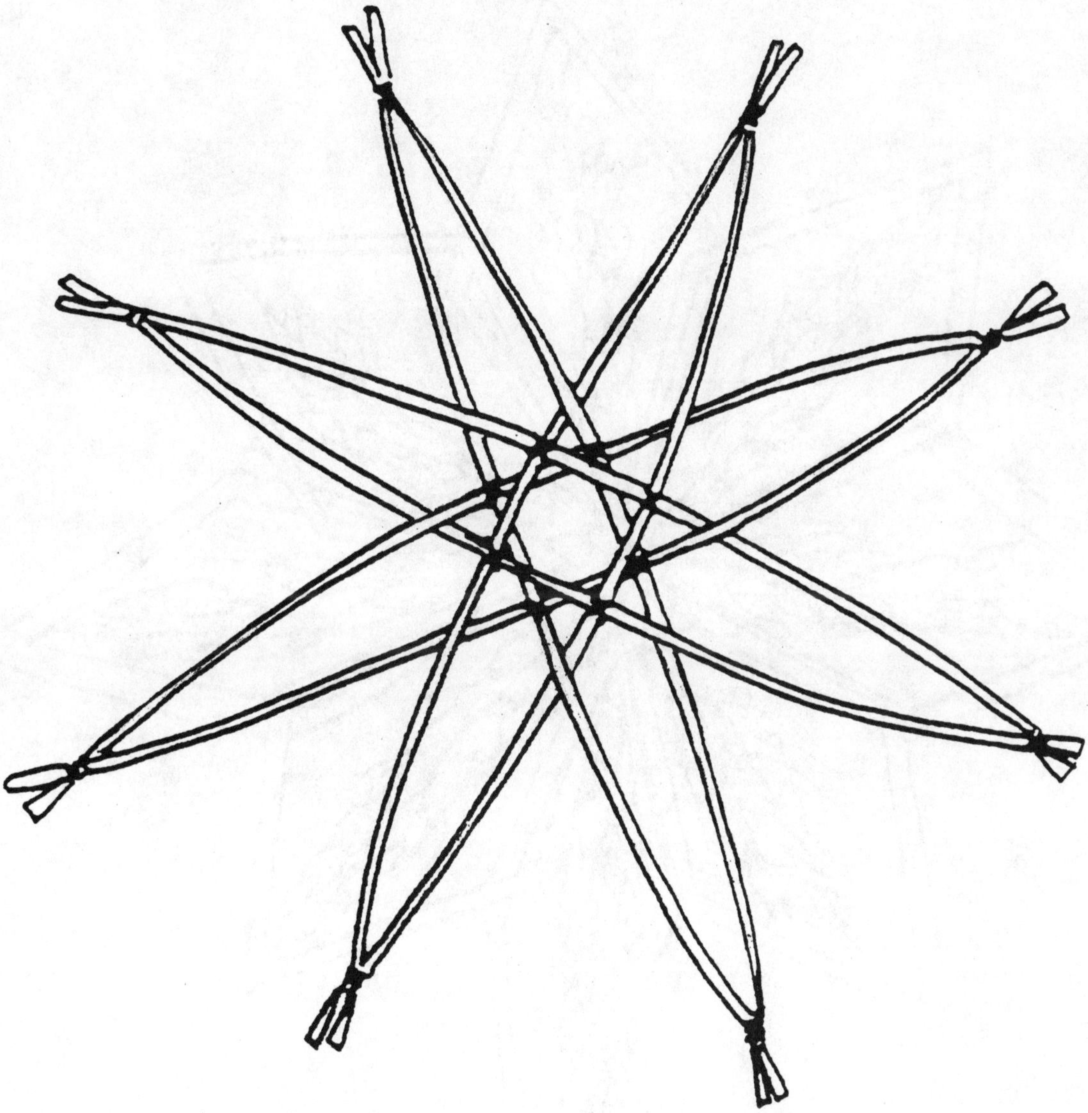

Diamond Star

Tie eight wheat stems at the necks. Fold the stems to form a diamond shape and tie again at the center. A thin wire may be inserted into the stems to form the shape.

Swedish Star

1. Make two wheat straw triangles.
2. Intertwine before attaching the ends of the second triangle.
3. Secure the triangles in place with a touch of glue.

Eight Pointed Star Snowflake

1. Select sixteen hollow wheat straws, about eight inches long, with uniform widths. Pre-soak according to the directions on page 17.
2. Tie all the straws together at the ends of the wheat stems.
3. Place all the wheat straws on a solid surface and form a circle with the stems.
4. Combining every two straws, from the center, tie, measure two inches, and fold wheat over. Measure one inch more and again fold the wheat stem over. Measure another two inches and tie the stems together at this final measurement.
5. Repeat all the above steps (#4) until all the straws are tied together in twos. When finished, there will be eight stem bundles of two.
6. Pull each of the stems apart and tie to its neighbor.
7. Continue to tie all the wheat stems.

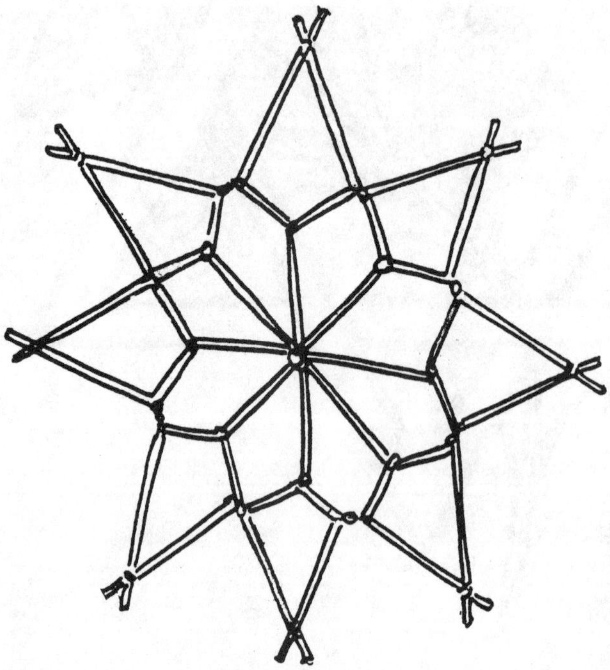

Chain Of Stars

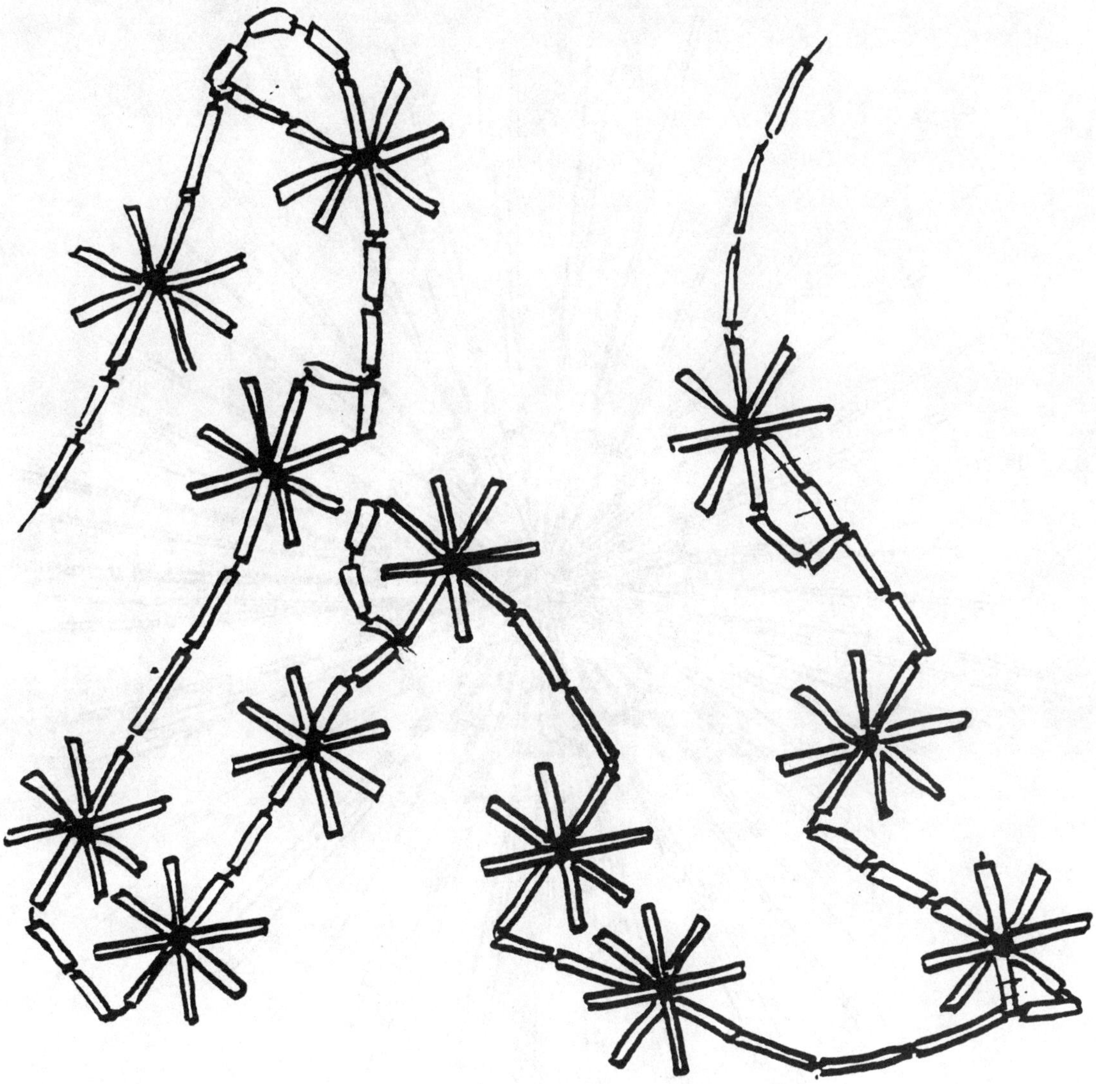

Run a needle and thread through 1/2" long pieces of wheat stems and tiny wheat stars. Use to decorate the Christmas tree.

Christmas Star

Christmas Star

Wheat heads on the ends of stems gives an entirely different look.

1. Decide on the number of stems to be used. Presoak according to the directions on page 17.
2. Tie all wheat stems together with the clove hitch knot at the neck. Place on a flat surface, fold the stems back to form a circle.
3. Approximately 1-1/2" from the center, tie, weave under, then over, each individual stem. Natural colored raffia is a good reinforcement. Return and weave in the opposite direction.

Snowflake

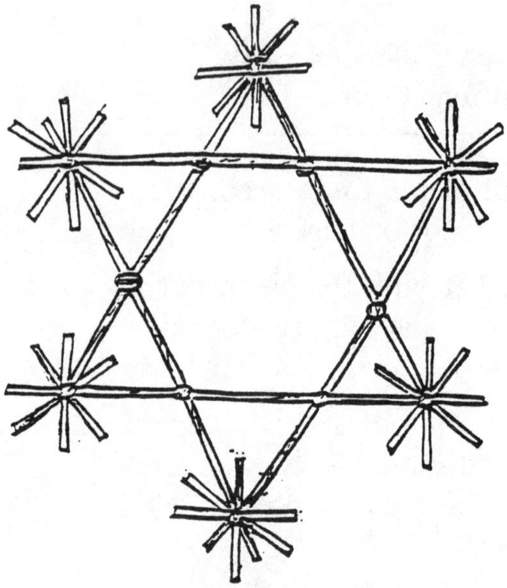

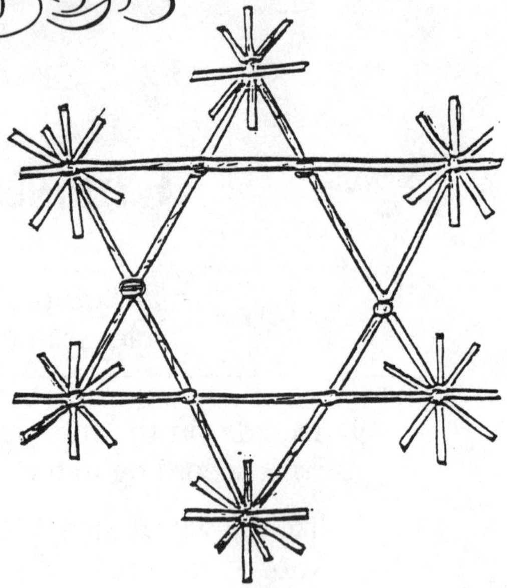

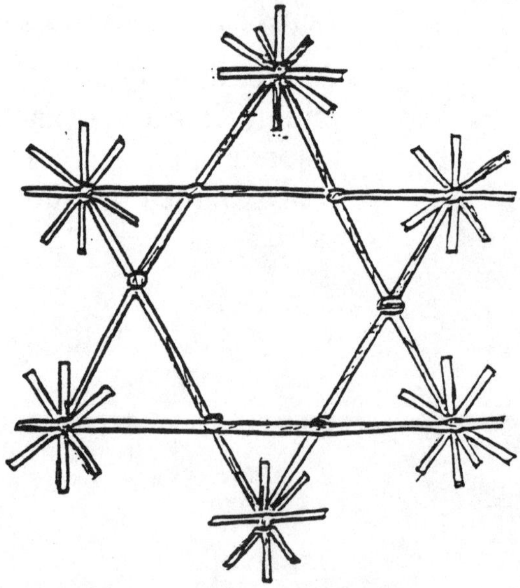

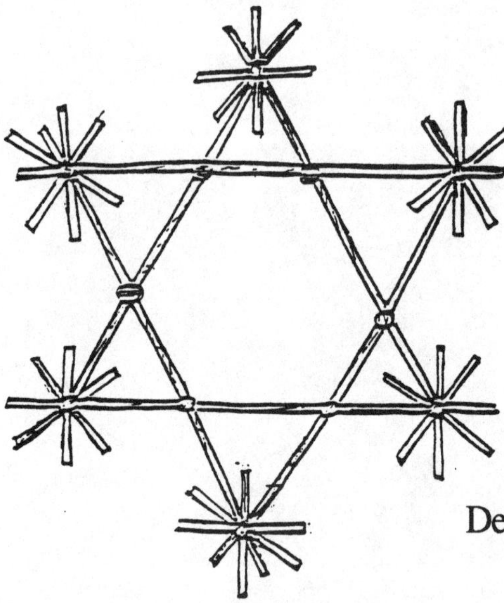

Delightful when hung on Christmas trees.

Tie two triangles together to make a six pointed star. Add six tiny stars which have been made earlier, with hot glue. Attach the tiny stars to the six triangle points.

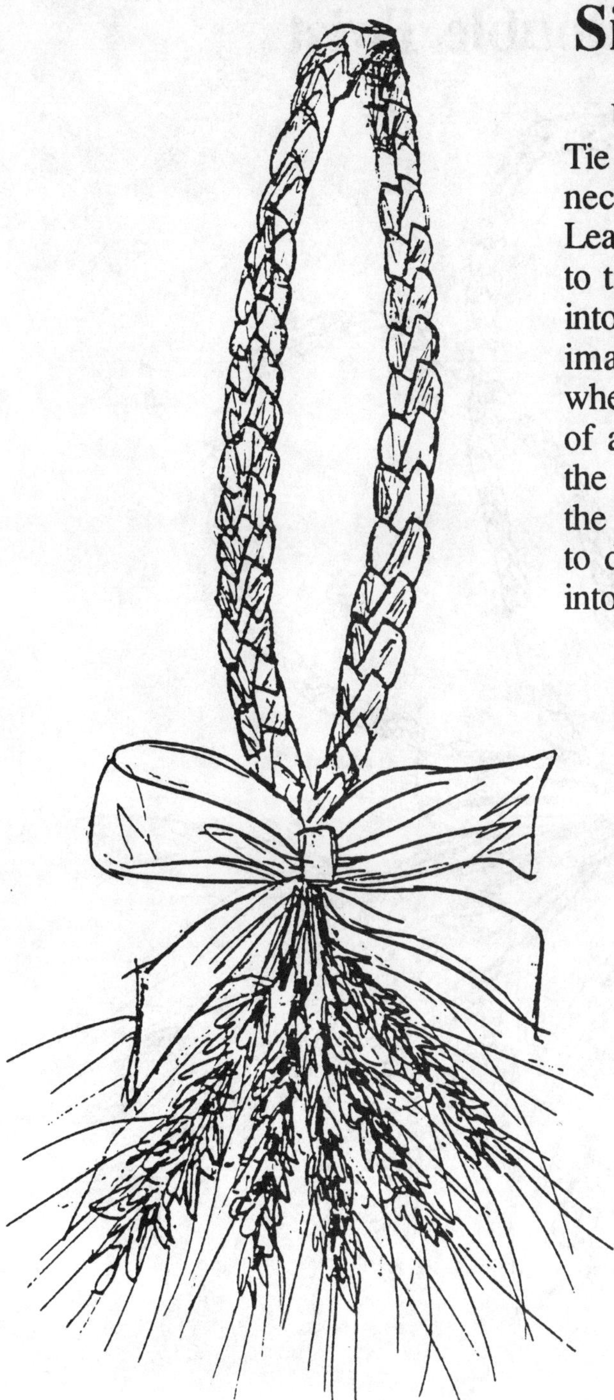

Single Large Twist

Use six stems of wheat. Soak. Tie the wheat heads together at the neck, using the clove hitch knot. Leave the string ends long enough to tie again later. Divide the wheat into thirds, then braid it for approximately 8-1/2 inches. Flatten the wheat stems out with the dull edge of a table knife as you braid. Tie the end of the braid to the neck of the wheat heads. Allow your work to dry with the wheat head spread into a fan shape. Add ribbon.

Large Double Twist

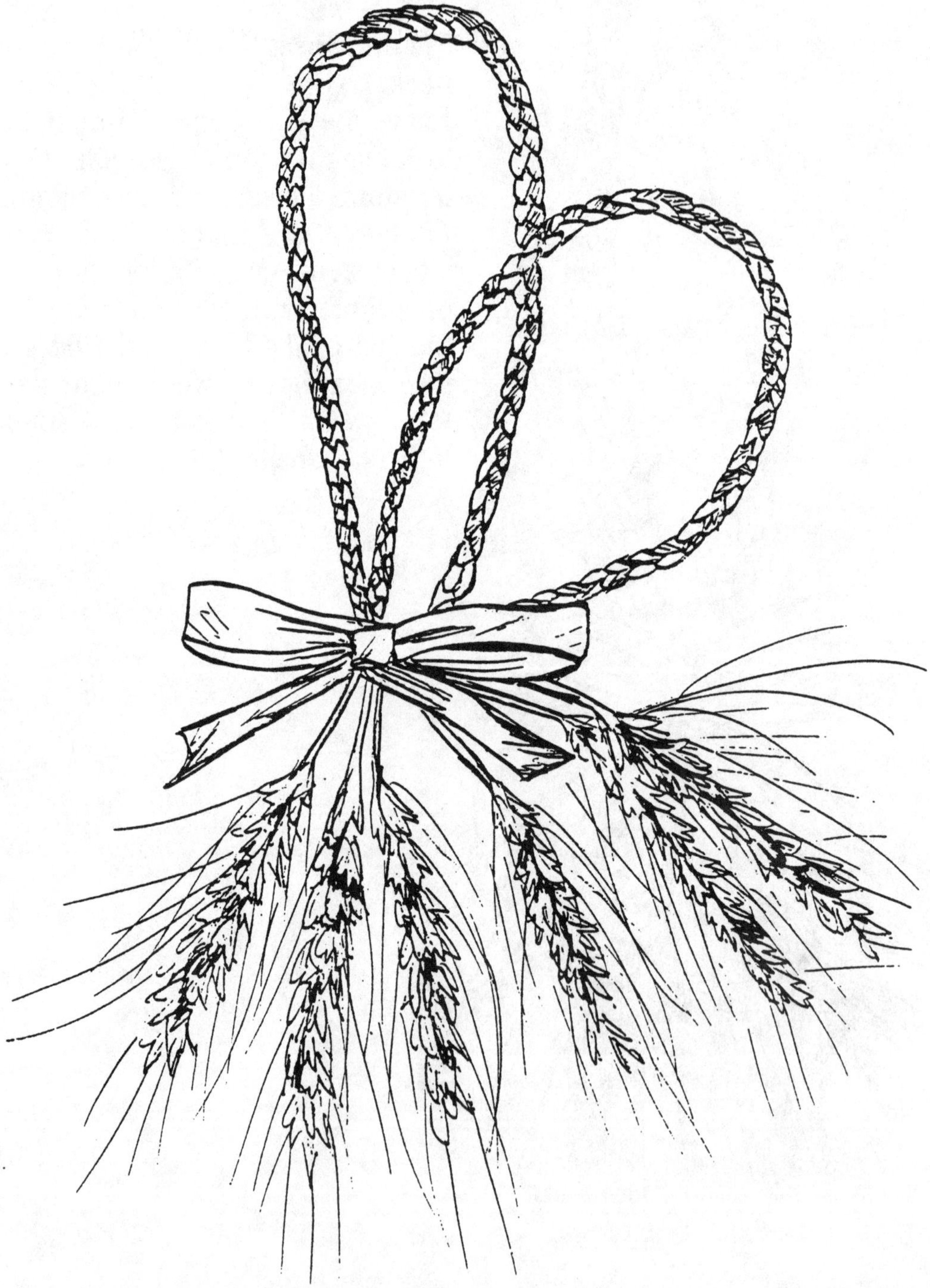

Large Double Twist

1. Choose six wheat stems with heads that are uniform in size and pre-soak according to the instructions given on page 17.
2. Divide the six stems in half and tie each set of three wheat stems together with the clove hitch knot.
3. A helper is needed at this point. (If a person is unavailable, consider pounding a large nail into a work bench with a third of the workspace available. This will secure the end of your plait. The tautness will give your plait a straight, even appearance.
4. Plait to the desired length and tie both the ends and their necks together. Flatten out and dry between two sheets of newspaper, weighted down by bricks.
5. When completely dry, tie a colorful ribbon on your Wheat Dolly.

Note: The more wheat stems added, the wider and thicker the plait will be. The flatter and more parallel the wheat stems are, the wider the plait becomes.

All wheat twists: small single, large single, small double, large double, small triple, and large triple are made according to the above directions.

Triple Twist

Small Double Twist

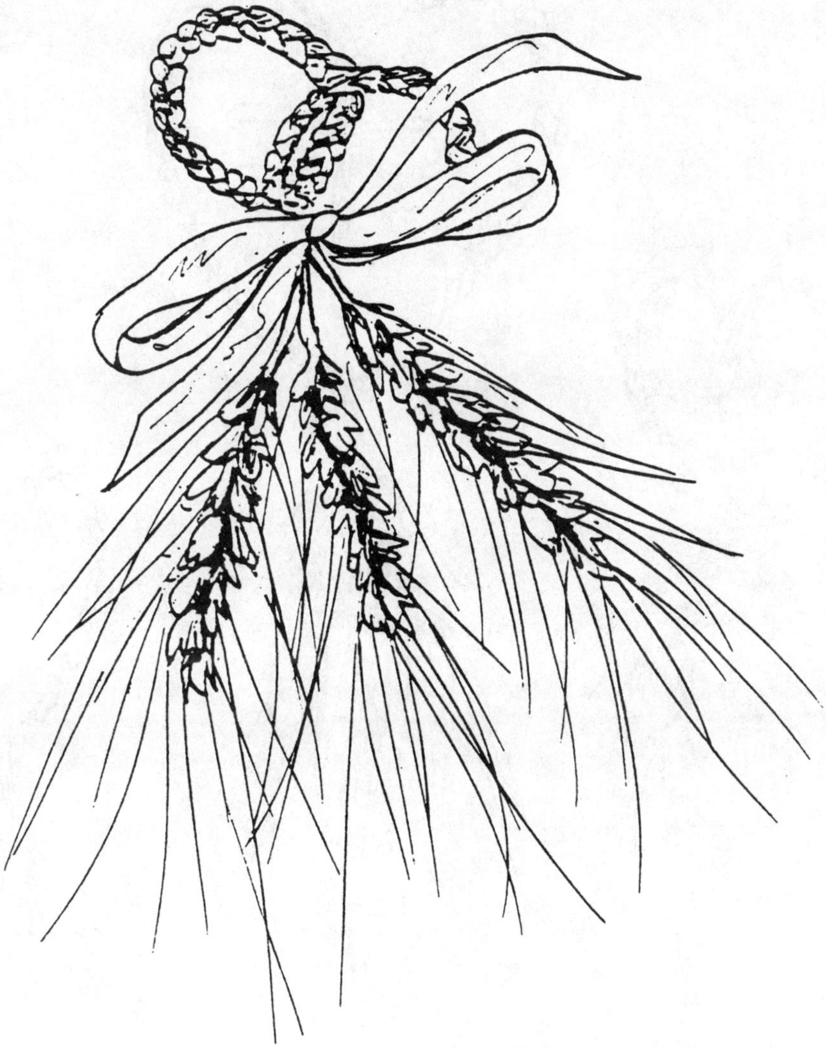

This simple beginner bow is made by tieing two three-straw plaits which form 2 circles together at their necks and wheat stem end.

Use directions for Large Double Twist, only shorten braid lengths to 3 1/2" long and then tie together with wheat heads and proceed the same.

Miniature Heart

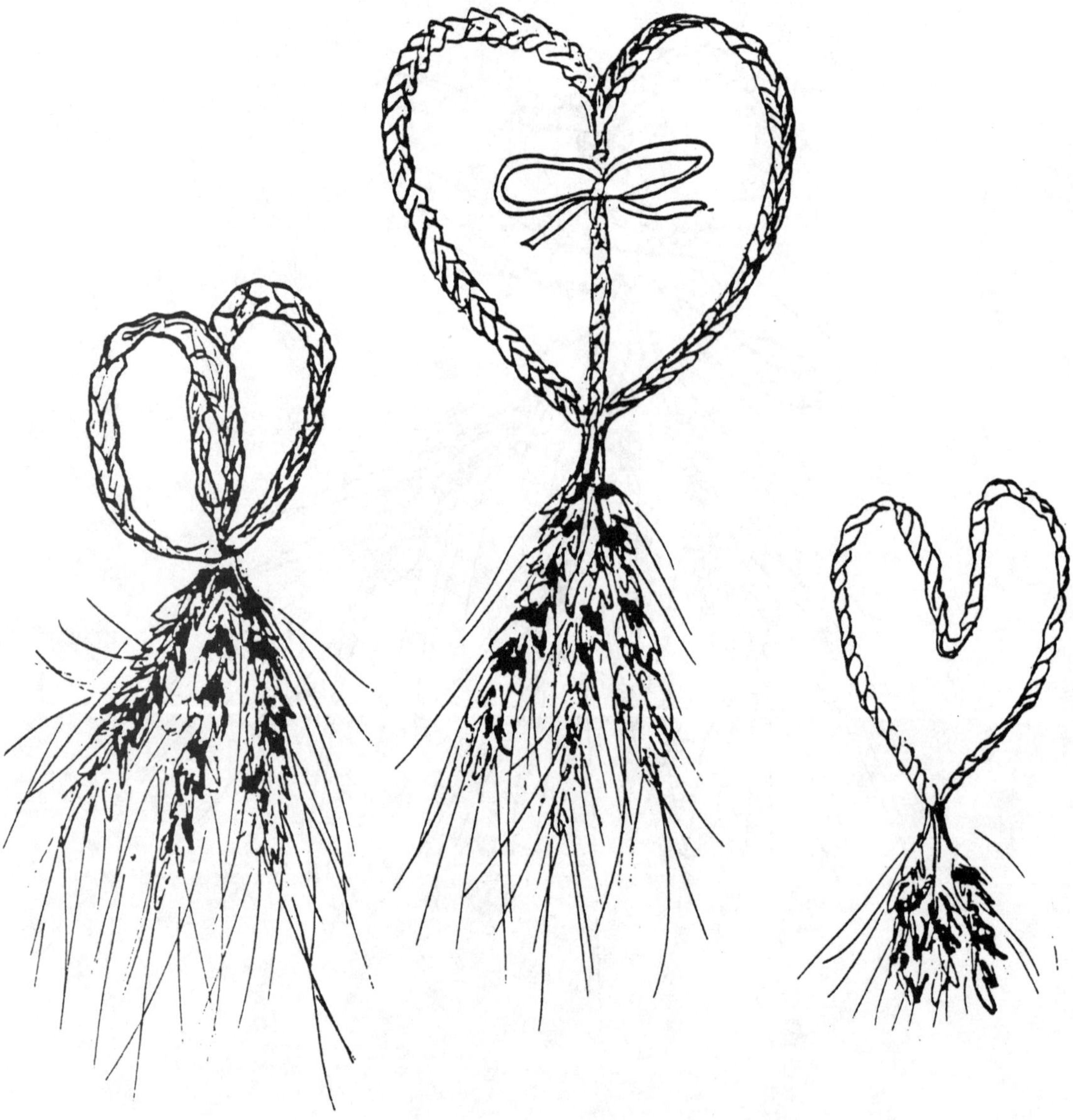

The Miniature Heart is for any occasion where a corsage is needed – Mother's Day, birthdays, weddings, anniversaries, dances, etc. – can be made with a fresh or dried flower attached.

Double Braided Heart

For that special person and that beautiful day!

Double Braided Heart

1. Soak 12 stems of wheat.
2. Tie three stems together at wheat heads. Fix securely, then braid, using only the flexible part of the wheat stem, adding more wheat stems as needed.

insert additional wheat stem here

3. Continue to braid to the desired length of the large heart.
4. Braid smaller heart and tie all ends together with clove hitch knot.

Step 4.

5. Add ribbon. Shape heart by bending in the center of each circle and forming a point by pressing ends together.

This design may also have a center wheat base. The design may use several other plaitings.

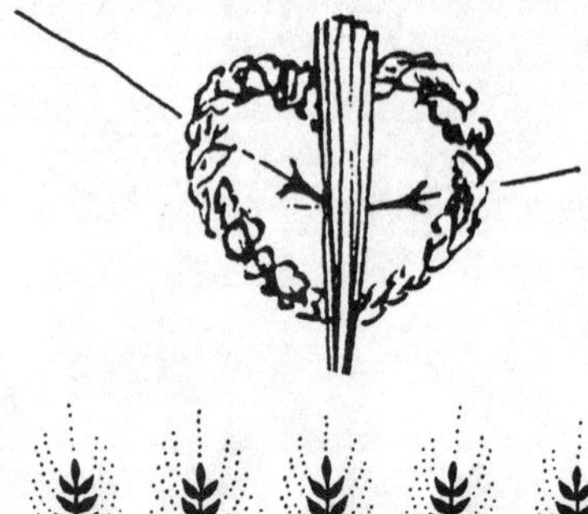

center wheat base

Figure Eight

Figure Eight

There is no wrong way to design an original Wheat Dolly. Everyone will be successful interpreting a style of their own. Even the beginner has an opportunity to become more creative and imaginative in their work.

1. Plait the desired length your Figure Eight is to be.
2. Tie to the upright center bundle of wheat.
3. Dry completely. Add ribbons to the top and bottom of the Wheat Dolly. Make a small loop at the back side to use as a hanger.

Any of the preferred plaiting may be used to your desired size formed into a figure eight and attached to a center wheat bundle of any number of wheat stems.

Gem

Gem

The Gem is similar in style to the "first" Corn Dolly I bought in Wales, England in 1971. Our company had great success with this charming Wheat Dolly.

1. Select nine wheat straws of uniform width with large heads. Pre-soak according to the instructions on page 17.
2. Tie all the wheat necks together using the clove hitch knot.
3. Divide the stems into thirds and flat braid. Flatten the stems as you braid and fold each braid over the dull side of a dinner knife. Braid half-way up – with all stems – then fold the stems over and place in the back of the braid. Tie all stems together at the neck.
4. Spread the wheat heads out and place them between newspapers. Dry in the full sun. When completely dry, add a colorful ribbon.

Tied Wheat Cross

Tied Wheat Cross

This is a wonderful gift to give at Easter time.

1. Select sixty wheat straws with large full heads attached. Pre-soak according to the instructions given on page 17. Decide on the size you want to make the "Cross" and cut the straw accordingly.
2. Divide the wheat straws in half. The first thirty stems will form the vertical part of the Cross. Tie this bundle of thirty at the neck and also at the wheat end.
3. Divide the second bundle of thirty straws in half. Fifteen wheat straws will go in one direction, while the other fifteen will go the opposite direction. The wheat heads and ends of the wheat will be intermingled. This will form the horizontal part of the Cross.
4. Raffia is a good choice with which to wrap and tie the Cross.

Jubilee

Jubilee

1. Soak 25 wheat stems.
2. Tie wheat stems together at neck of wheat.
3. Bend 5" back.

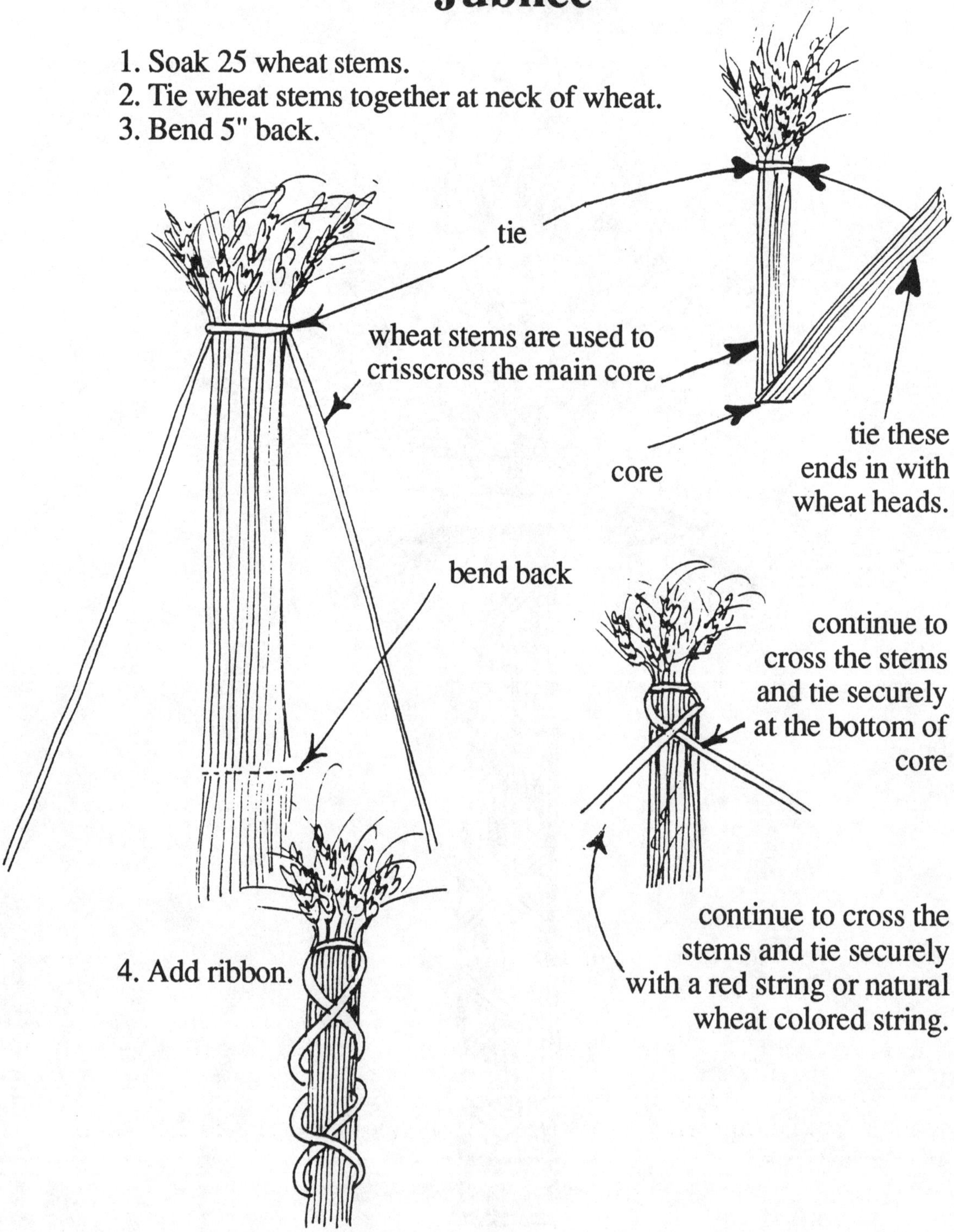

4. Add ribbon.

Harvest

Use 25 stems of wheat. Tie all wheat heads securely at the necks, using clove hitch knot. Holding wheat heads with one hand, with the other hand twist all the stems over in the opposite direction. Keep twisting and turning until a twisted form has been made, about the size of being bent around a silver dollar when the stems form a circle. Tie securely, being sure the twist is not allowed to unwind. Trim the stems at a nice angle. Spread wheat heads out and allow to dry. Add a colorful ribbon if desired.

Napkin Ring

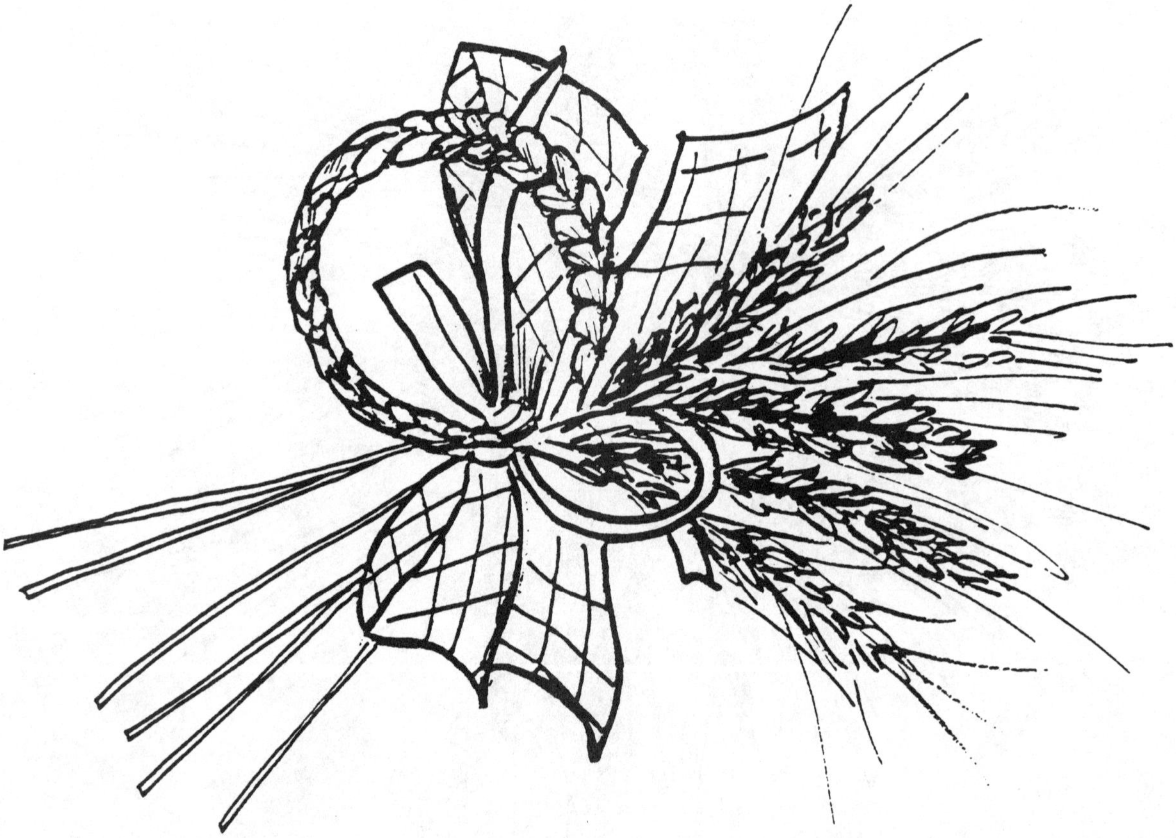

Use nine stems of wheat. Soak in warm water. Tie all wheat heads together at necks, using clove hitch knot. Leave ends on string which will be used to tie finished braid in place. Divide wheat into thirds. Braid, flattening the wheat as you braid. Press over the edge of a sharp dinner knife. The braid should be wide and flat. Make a 2" diameter circle with the braid, then tie the circle in place with ends of string. Trim ends of wheat , fan out, dry, and add ribbon.

Petit Sheaf And Circles

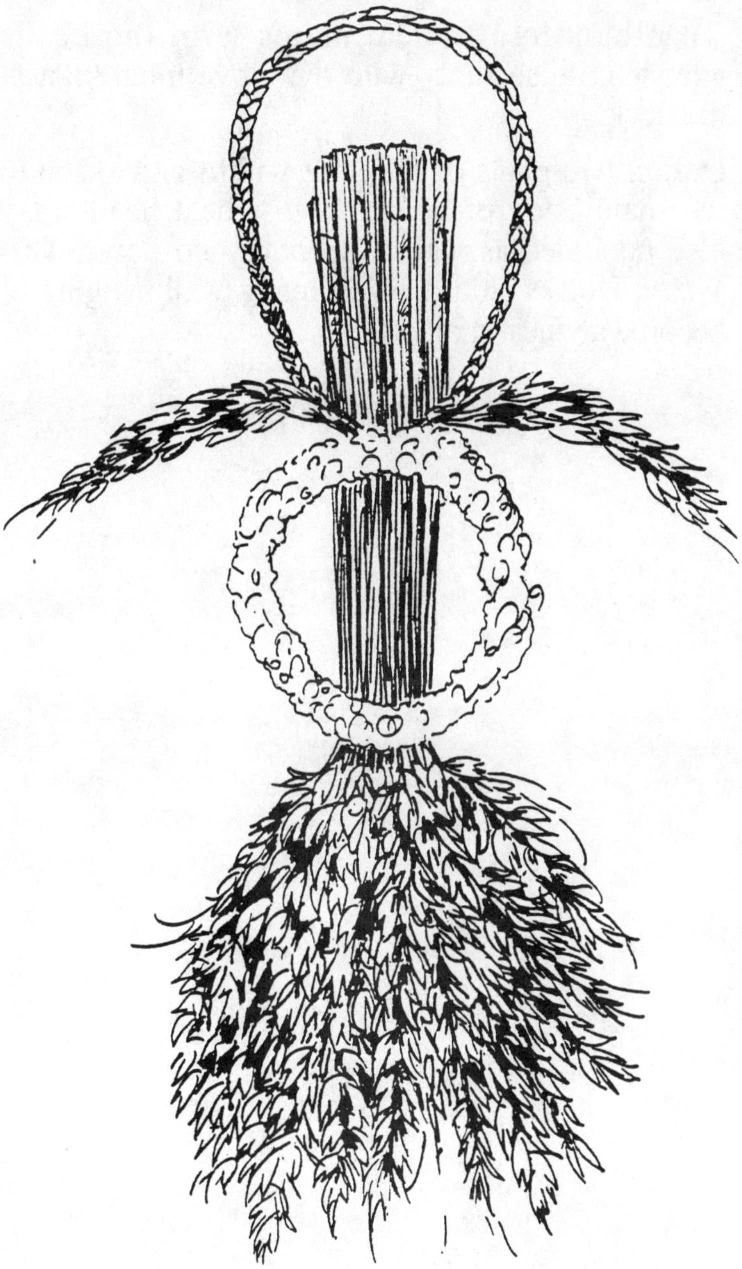

Petit Sheaf And Circles

1. To a bundle of wheat stems with the heads attached, tie securely with the clove hitch knot at the neck.
2. Design two plaits of different widths and attach to the main body of wheat. Two wheat heads may also be added as well as ribbons and bows. This Wheat Dolly could develop into a wall hanging or become a wheat maiden.

Four Straw Plaited Circles

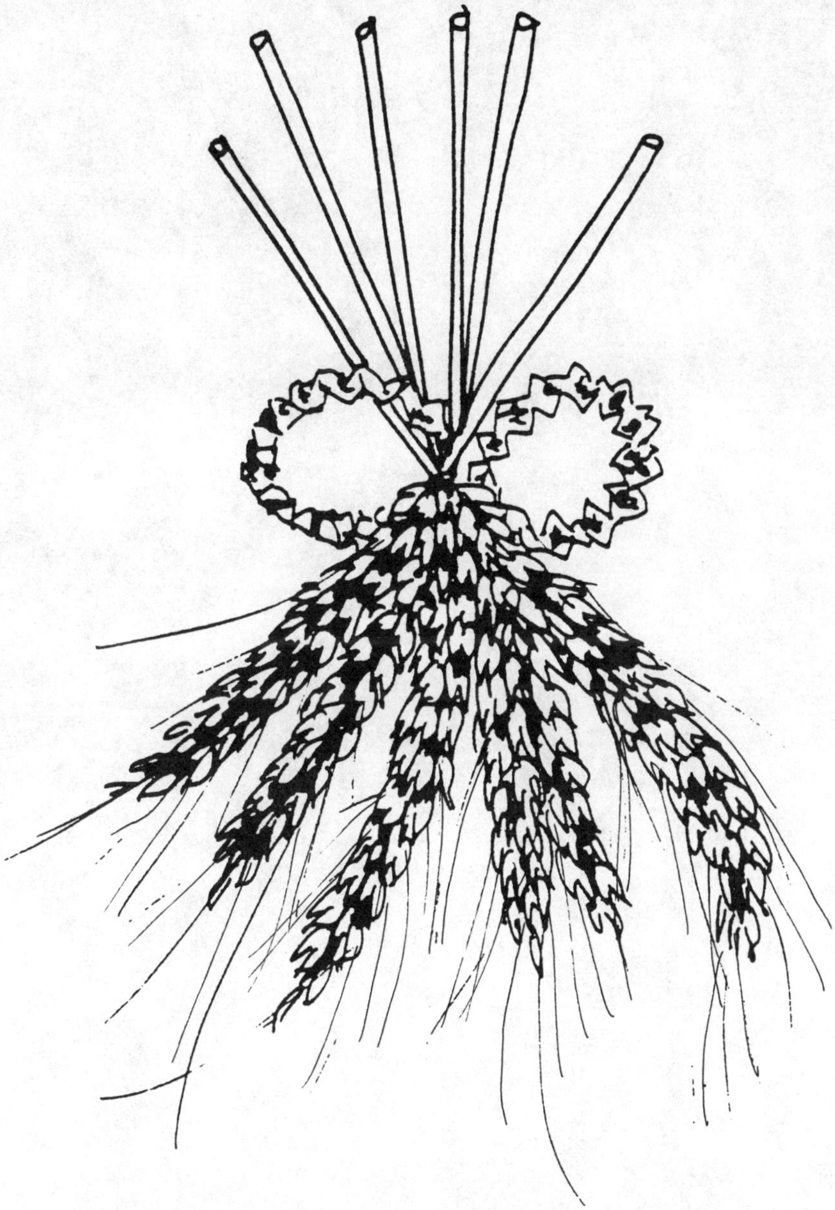

Star of David

Intermediate Wheat Dollies

CHAPTER
7

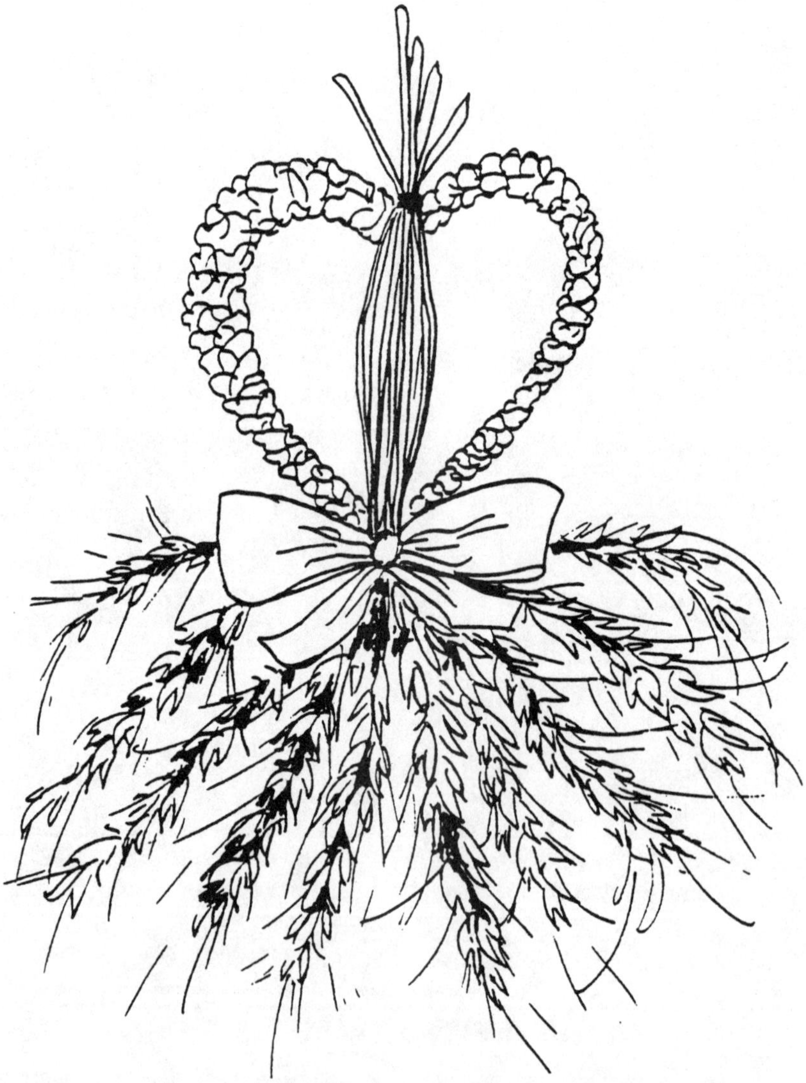

The Bell

Single Bell

A Five Spiral Plait

The Lantern, Pyramid, Bell, Dolls and Flowers and many other designs are made with the five straw plait. Begin with five wheat straws with wheat heads. Plait one inch of narrow spiral. Turn upside down (the heads will point upwards). Start plaiting from this side and widen the work (or increase) as you plait. When you have reached the neck of the heads, the sides will be the length of the wheat heads. Turn upside down again (the heads will point downwards). Plait upwards. Increase the size the first inch and a half, then gradually decrease until the top of the bell has been reached. At this time a ribbon hanger may be secured to the bell, or a narrow spiral may be plaited as the hanger.

Double Bells

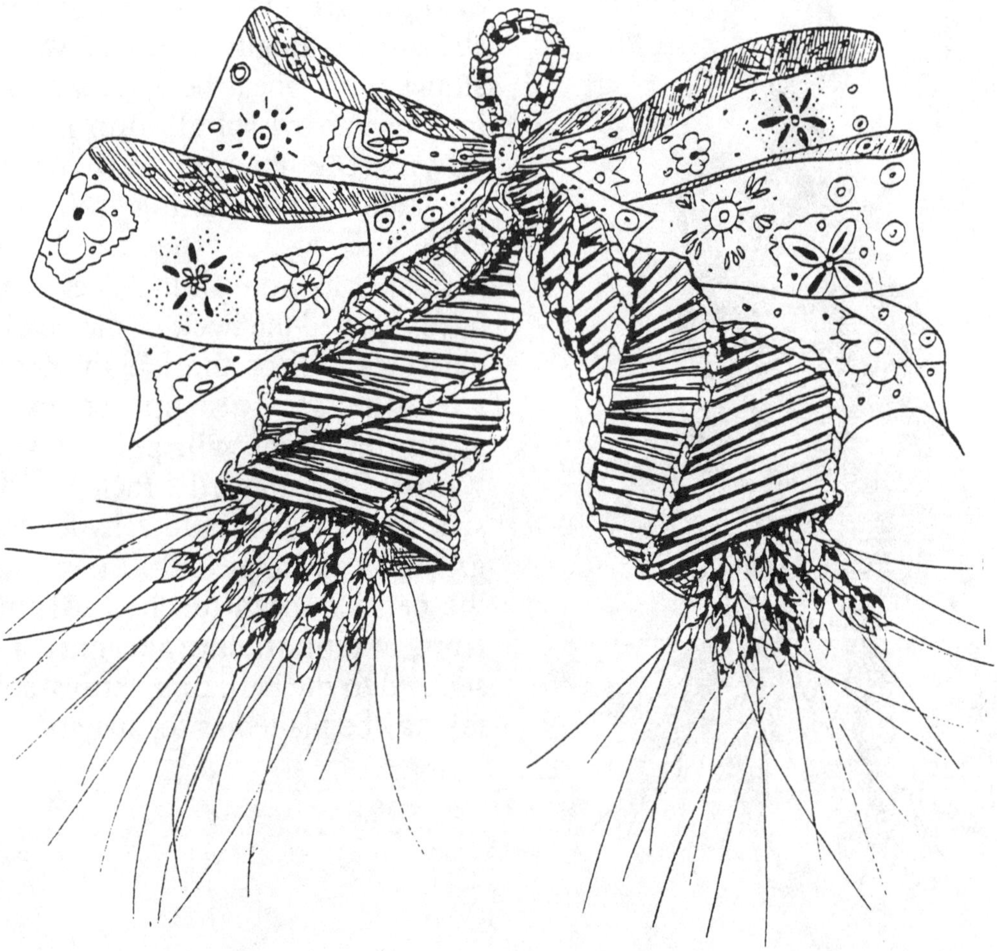

Wildrose Wheat Fan

(An Old Welsh Design)

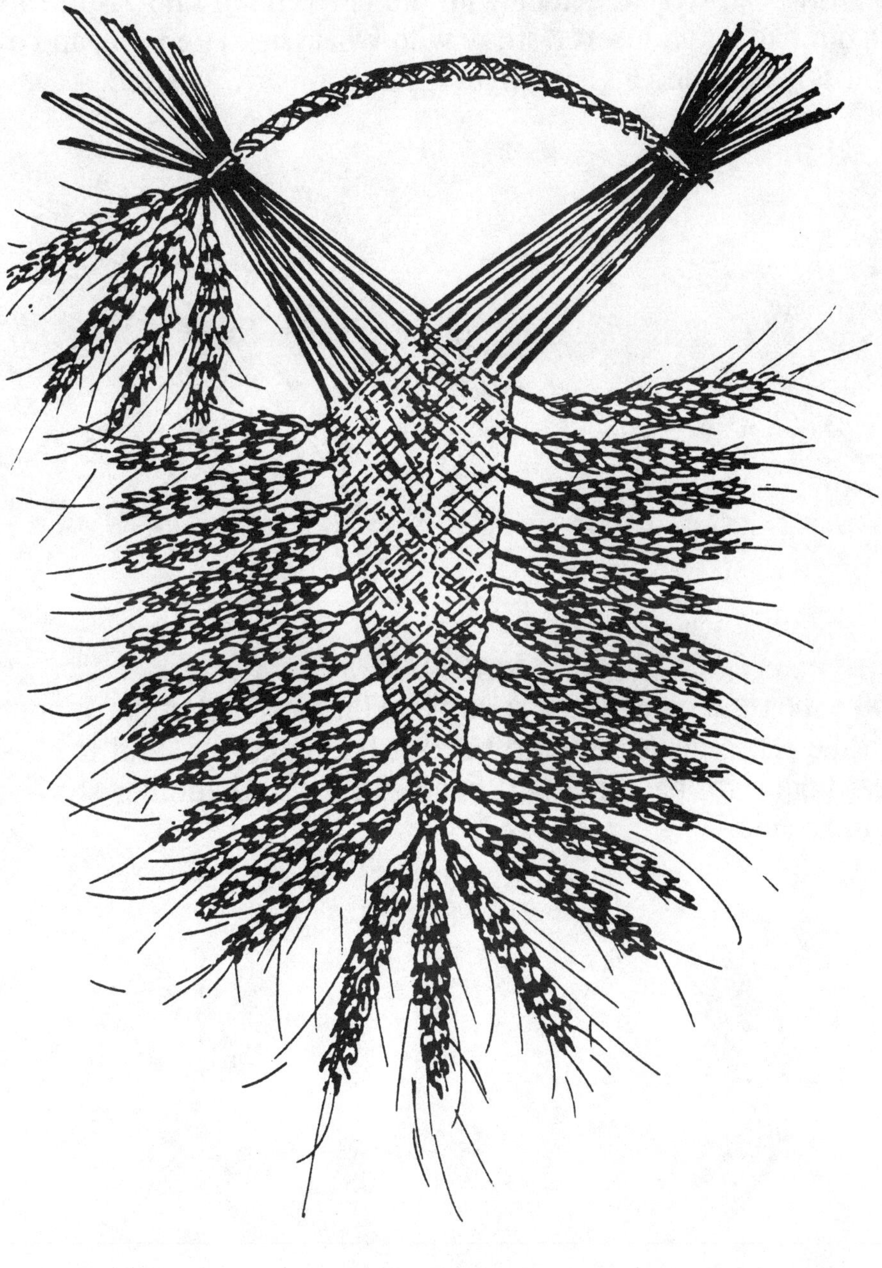

Wildrose Wheat Fan

The Wildrose Wheat Fan, an old Welsh design, is made with 25 straws of wheat with large well formed wheat heads. Begin by tying three straws securely together with the clove hitch knot. Spread wheat heads out like a fan. Insert a straw with wheat head under A and over B. Then insert stem under C and over the one already inserted.

1.

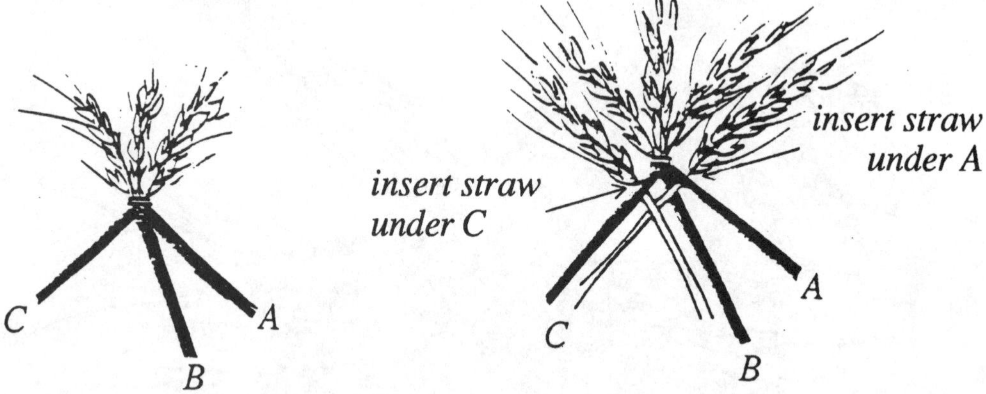

Take the right outer straw A and bend it back and under B and to the left. Take left outer straw and bend it back under C and to the right. Repeat both right hand and left hand straws. Add another straw on the right hand side.

2.

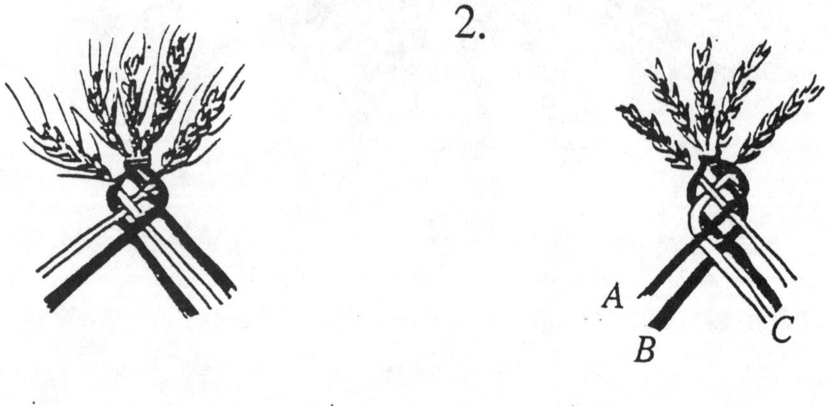

3.

4.

Adding more straws.

5.

Finishing Wheat Dolly

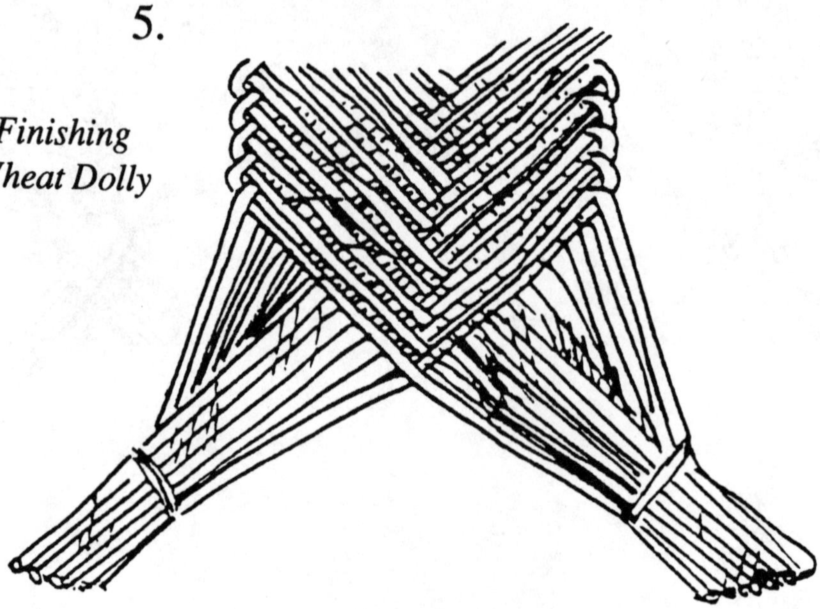

Welsh Fan

The Welsh Fan has many wheat dollies designed under this title and as many interesting ways of finishing them. The Welsh Fan is an excellent design for those new to plaiting. These are very nice as wall hangings. Design a more elaborate hanger. A simple, yet effective hanger is shown above.

The Long Fan

Follow the directions for the Welsh Fan. After 15 straws are used, cut off the third straw from the center on both sides. As long as new straws are added, continue to cut out the third straw from the center, on both sides. When only five straws remain, plait the handle, and tie securely in place. Trim off wheat stems. Remember, whenever new straws are added the fan becomes wider so it is necessary to cut out as many straw stems as you add. In this way you can make the Long Fan.

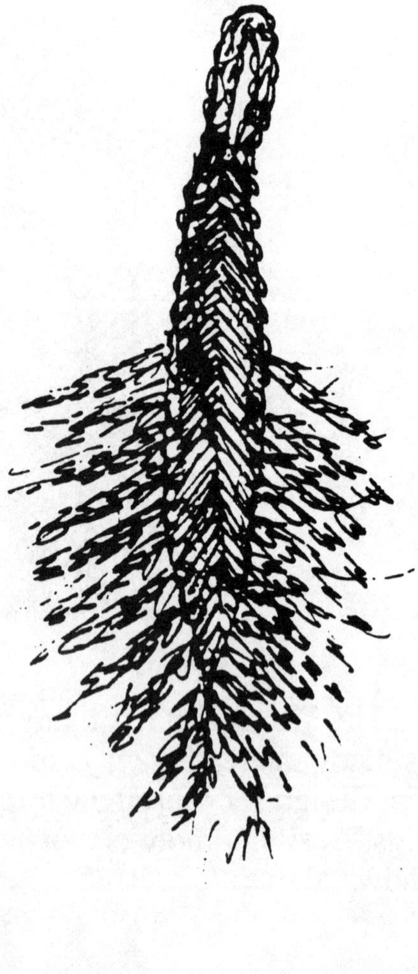

Christmas Tree

Follow the directions for the Welsh Fan only finish sooner than you would for the Long Fan and without the handle. Design your own tree base if you like. Attach a tiny gold wire hanger to the top of the tree!

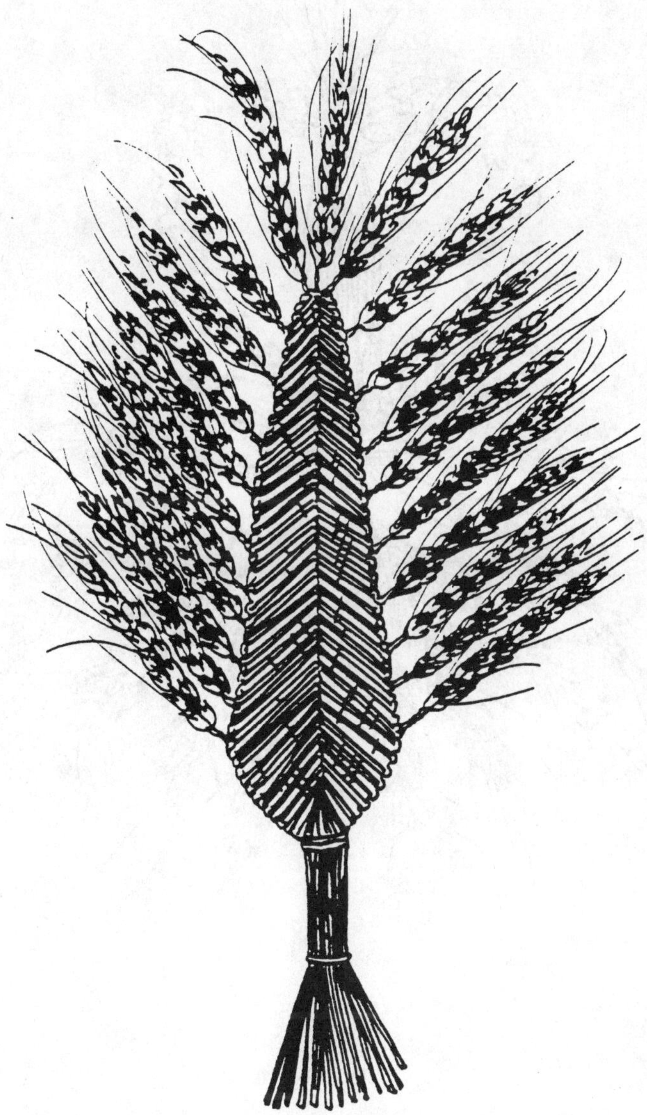

Mordiford

Single Mordiford

1. Soak 25 straws with wheat heads.
2. Make two seven straw plaits 6" in length, leaving 3" of straw unplaited at beginning.

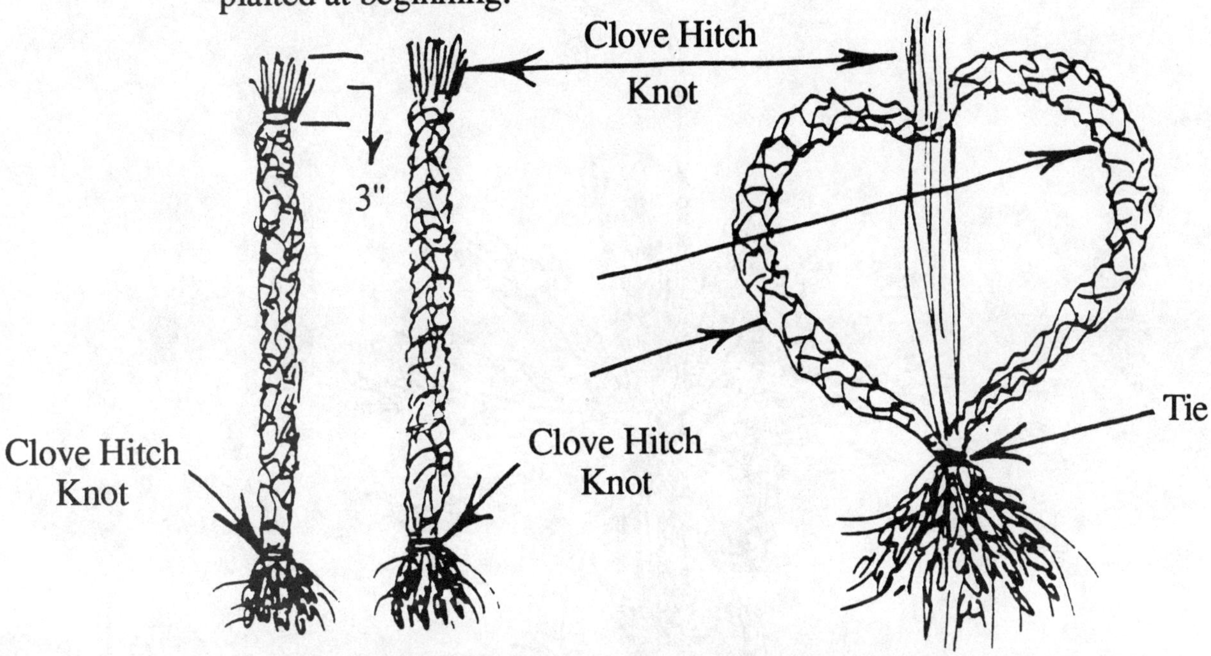

3. Tie wheat ends together using the clove hitch knot.
4. Now tie wheat heads together and form a heart shape.
5. Cut wheat ends off evenly at the top.
6. Secure with raffia.
7. Spread wheat heads out flat and dry in full sun.
8. When dry, tie colorful ribbon on form.

Double Mordiford

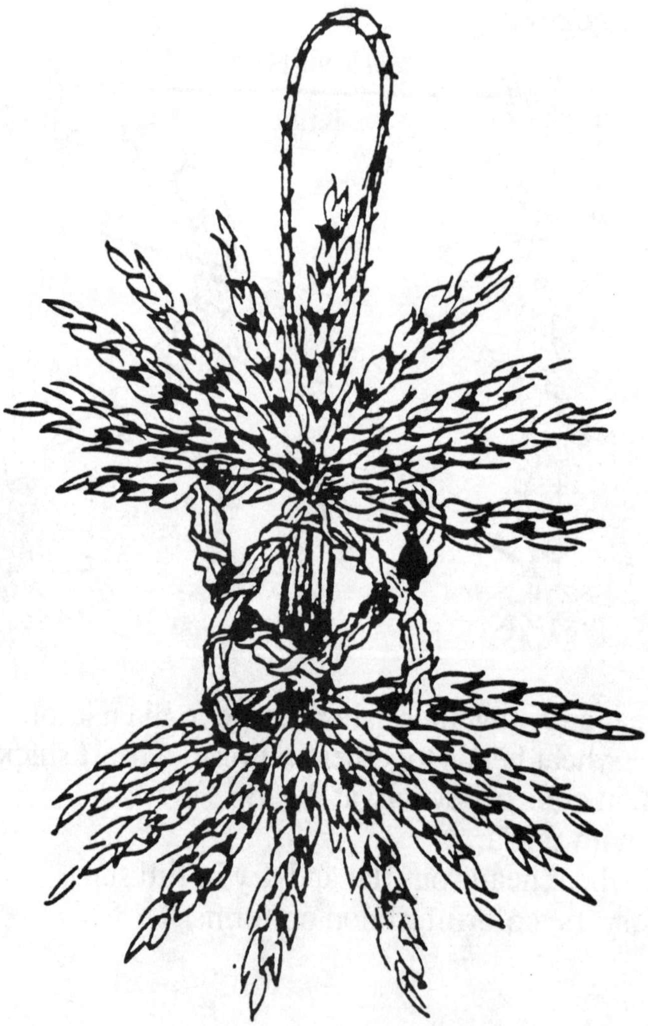

House Blessing

The House Blessing design is one of the loveliest wall hangings. It can be made in several sizes, but 12" to 14" seems to be ideal.

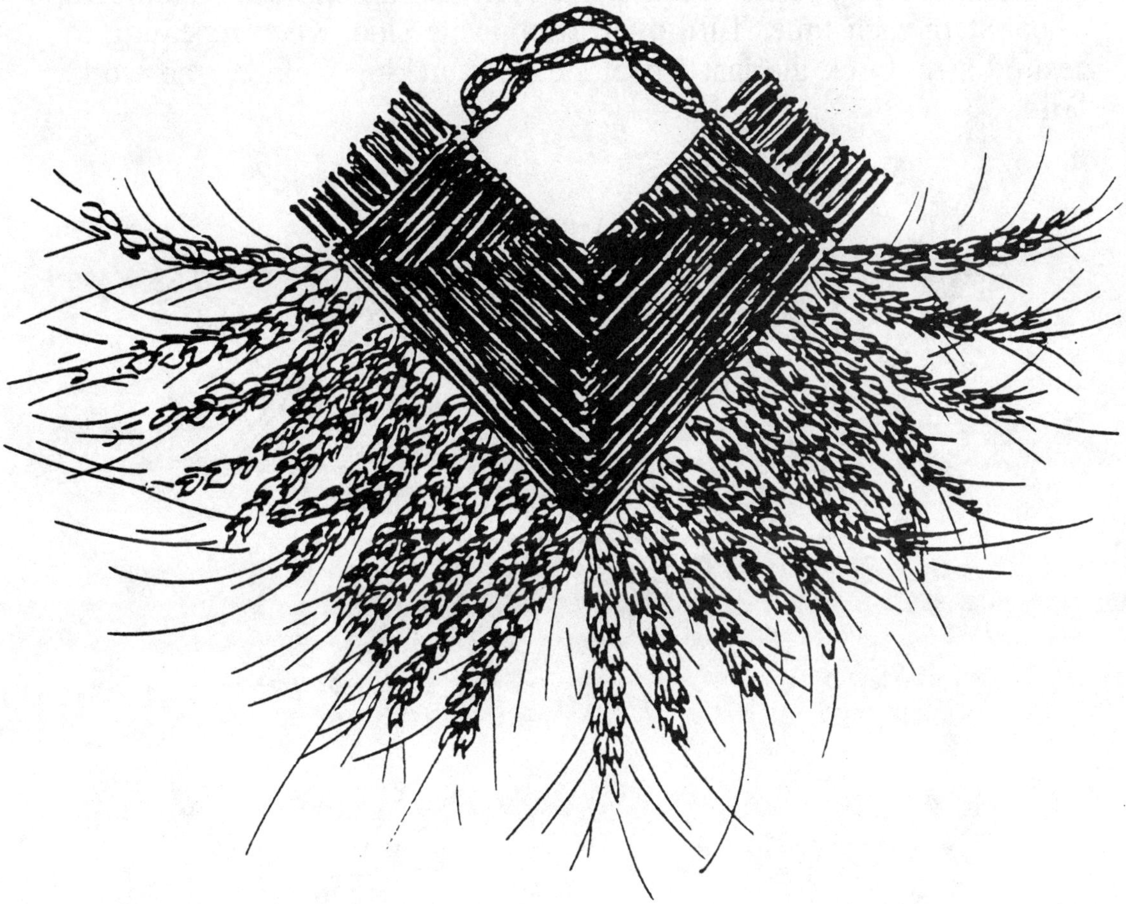

House Blessing

28 Straws

Cross the first two straws of wheat with heads about 1" from the neck. With a third straw, lay it next to one straw, then bend back over the stem of the opposite straw. Bend back and around each additional wheat stem each time. Turn over to opposite side. Keep repeating to desired size. Lock the last wheat stem by tucking it under the work. Trim.

Step 1

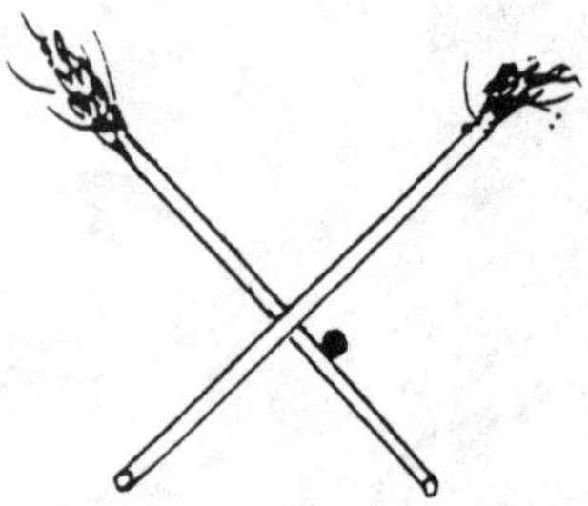

Step 2

Step 3

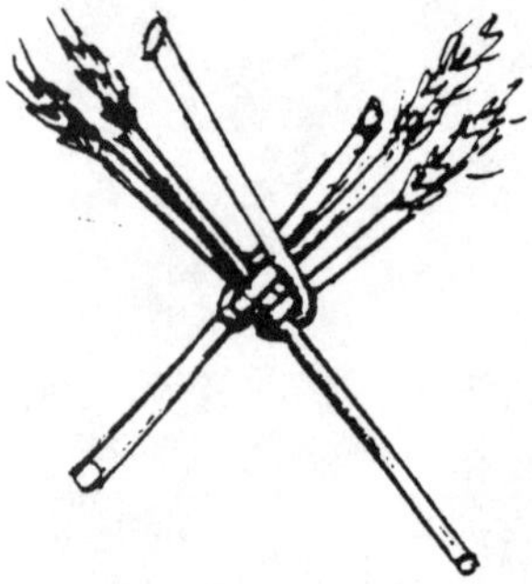

Step 4

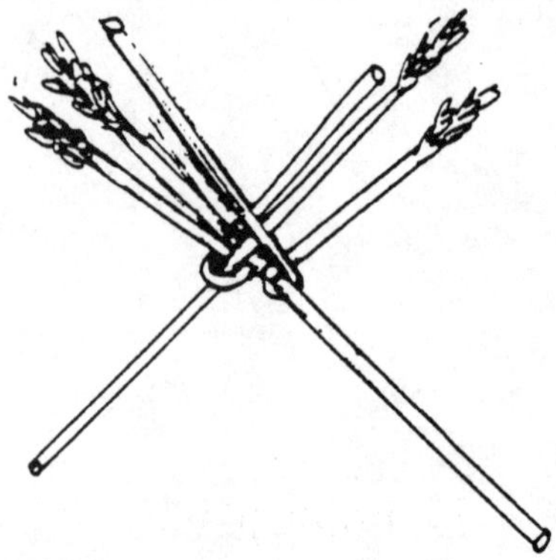

Sweetheart

A simple plaited heart with plaited hanger, ribbons and bows.

Wheat Spiral

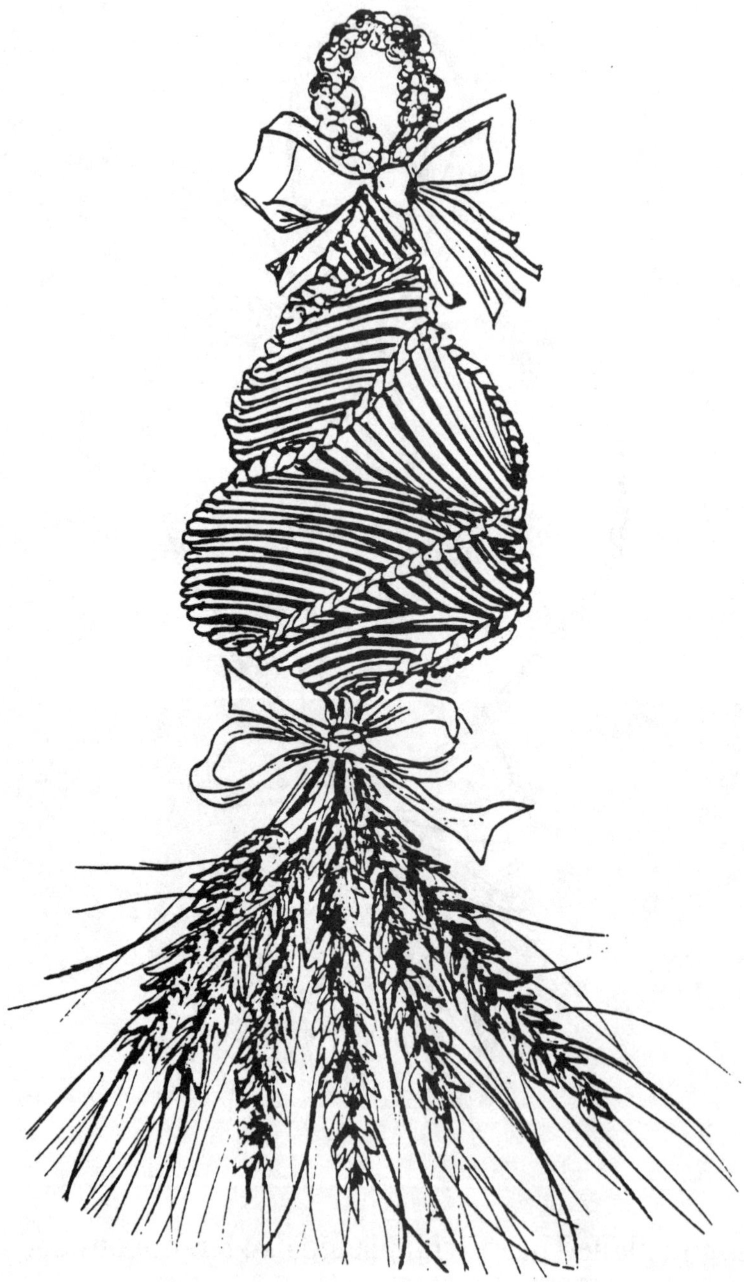

Wheat Spiral

The Wheat Spiral is hollow inside. It is made with the five straw plait, adding and decreasing where necessary to create its spiral shape. Spirals can be made in many sizes, from petit to large, as for a cornucopia.

Tiny - Petit Spiral

The Tiny Spiral is made with small diameter wheat, the thinnest, finest wheat you can find.

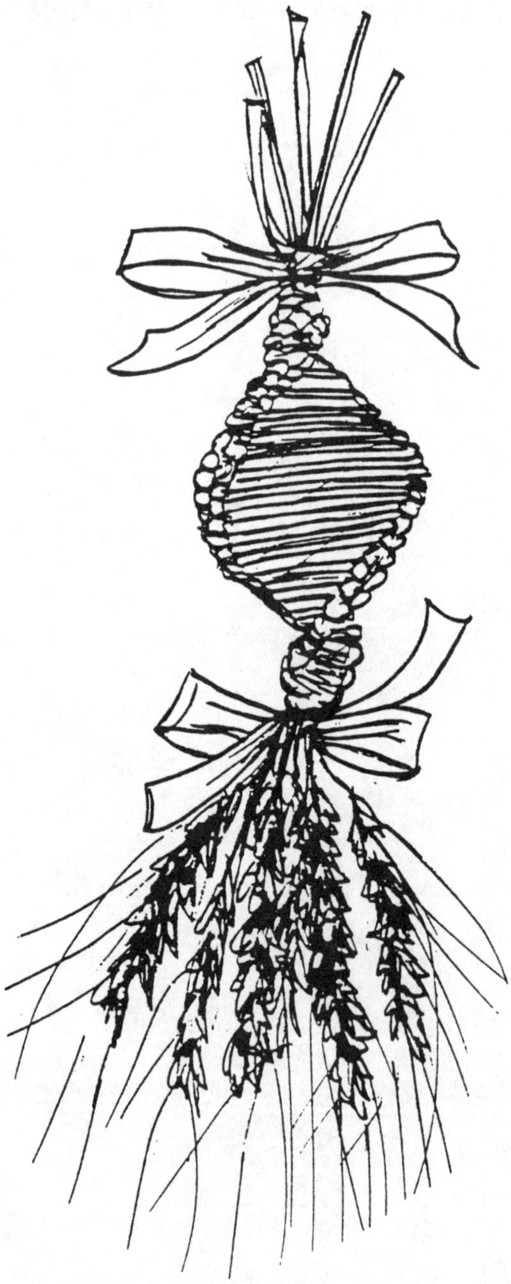

Hearts

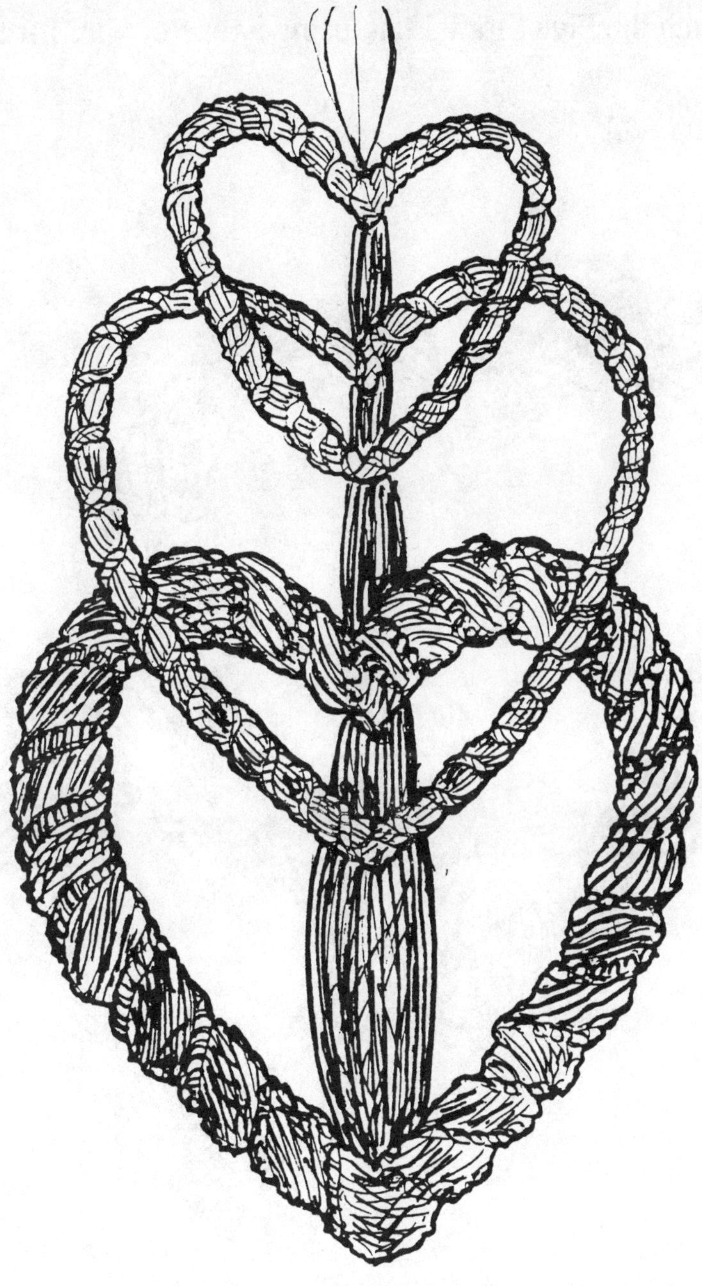

The four-straw plaited heart can be made with or without a core.

Hearts

Made with the Five Straw Plait, using a rope or wire for the core.

Flame

Five Straw Plait with rope core in oval and a wooden core in the candle.

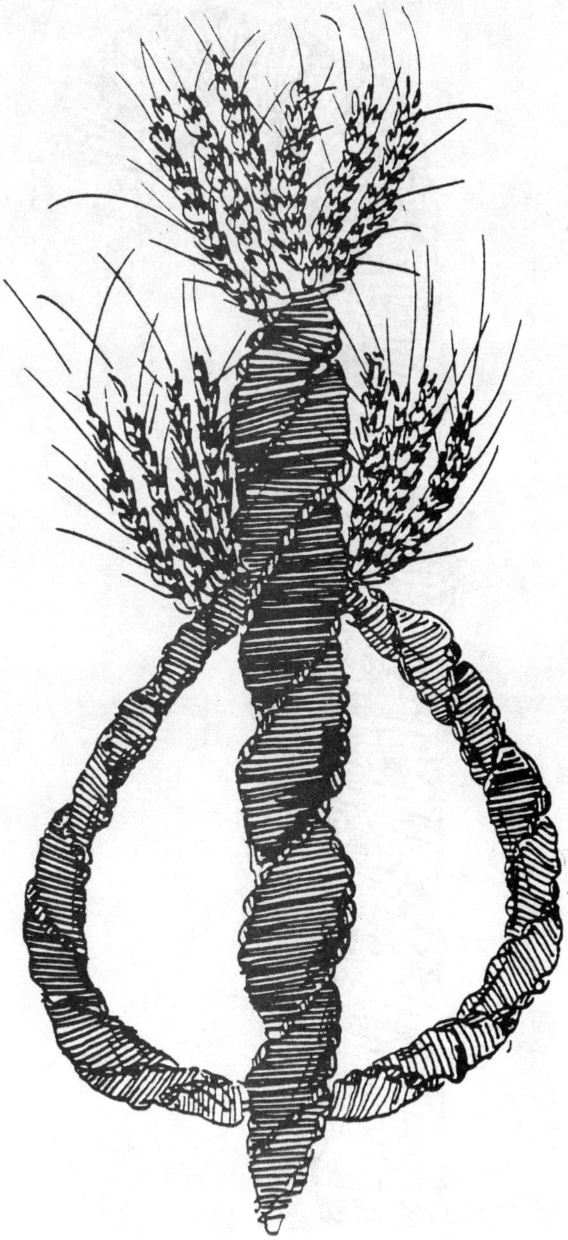

Wheat Cane for Christmas

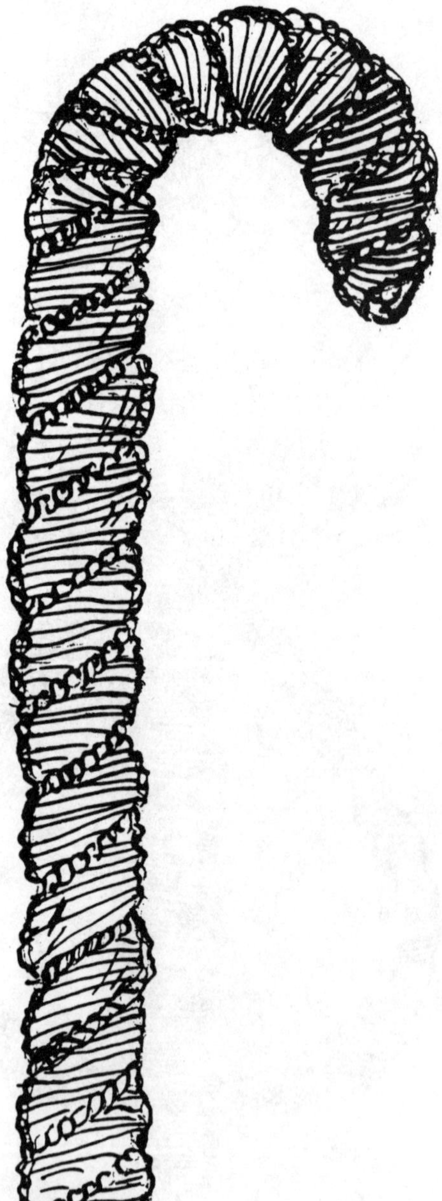

Suffolk Basket

Made with the Five Straw Plait. Increase to desired size, then decrease and plait a handle.

Umbrella

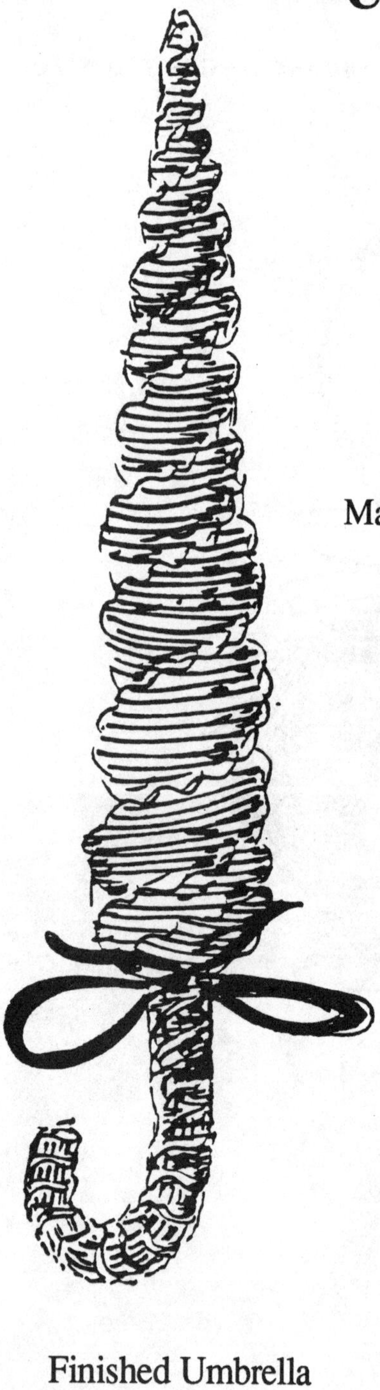

Finished Umbrella

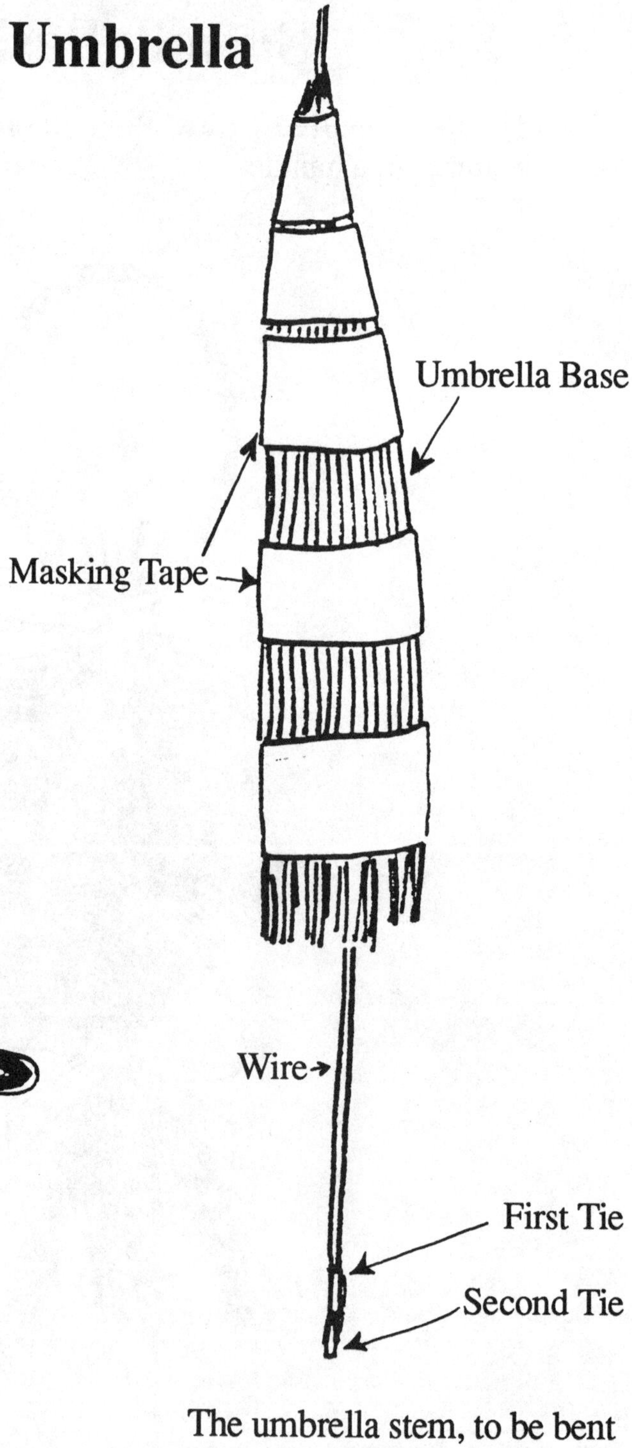

The umbrella stem, to be bent

The Umbrella

1. For the umbrella base, select 25 wheat stems that have not been pre-soaked. Tape these stems together onto an 18 gauge wire to form the shape of an umbrella, using masking tape to keep it all secure.
2. Begin plaiting at the handle.
3. The five straw plait is used with wheat that has been pre-soaked. Secure five wheat stems one and a half inches from the end of the wire handle, using the clove hitch knot. Tie again at the end of the wire, using the clove hitch knot.
4. Bend the wheat stems into a horizontal position.
5. Use the parallel plait to create the handle as far as the wheat base. Then begin the increasing plait to cover the base.
6. Continue to plait until the shape of the umbrella takes form. Begin to decrease the plait until the top of the umbrella has been reached.
7. Begin and end the umbrella with small diameter wheat stems, then gradually add the larger, thicker wheat stems at the widest area of the umbrella.
8. Finish the umbrella by tucking the working stem securely under the wheat stems.
9. Add a ribbon to the handle.

Heart Theme

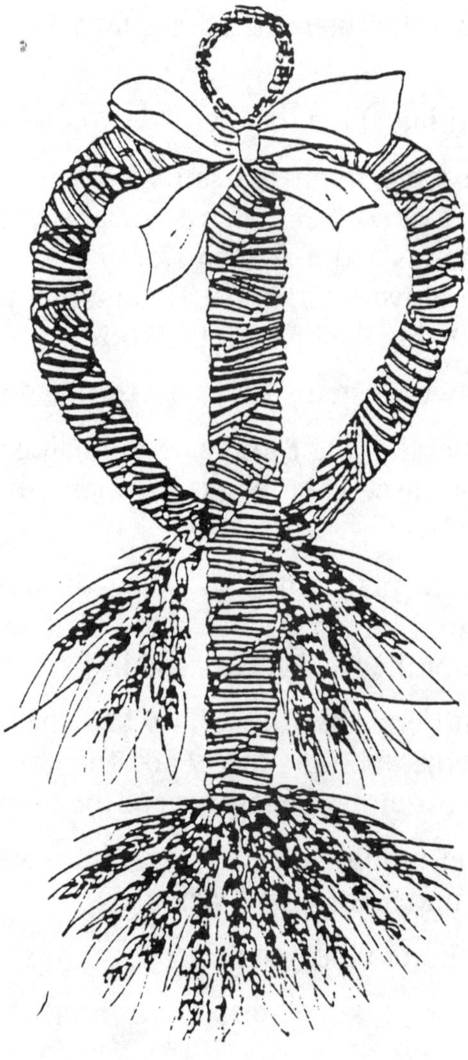

Two Three Straw Plaits

Advanced Wheat Dollies

CHAPTER 8

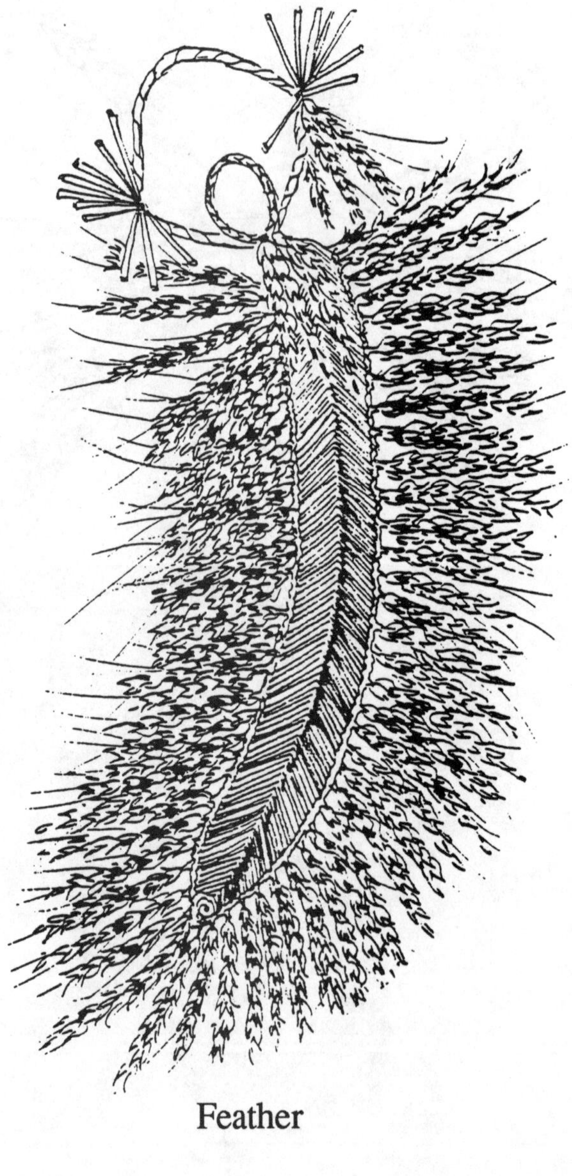

Feather

Feather

The Feather is made by the same principal as the Fan. Decide on the width you wish it to be and keep cutting out the straws as you work. When a straw is added, the design will stay uniform in width as you continue to cut straws out.

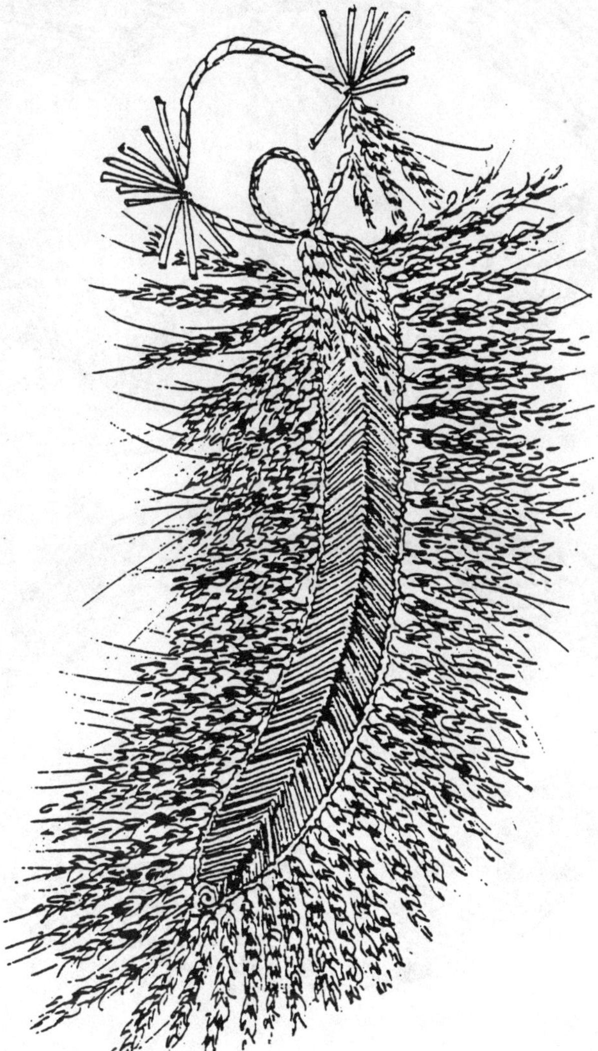

Wheel

The Wheel

1. The Wheel is made by plaiting around a rope. The ends of the rope are attached to form a circle.
2. Six Gems (without heads) become the spokes and are attached to the wheel. Use a needle and thread to attach them.
3. The hub-cap covers the centerspoke joinings. It is made from three plaits, which are arranged to form a six pointed star.

Handbell

Handbell

1. A large handbell is made with a nine-straw spiral plait.
2. The core for the handle is made from the wheat stems and the wheat heads become the clapper of the bell.
3. Select nine wheat straws with large full wheat heads. Decide on the size of the bell to be made and plan the handle accordingly.
4. Tie the wheat at the neck securely with the clove hitch knot. The rest of the wheat stems form the core of the handle.
5. After determining the length of the handle, tie the wheat stems again at the top of the handle.
6. Spread out the nine straws over this second tie and work a tight nine straw spiral plait back down over the core, to the end of the handle.
7. Now change the plaiting for that of a hollow spiral.
8. With handle pointed downwards, the plaiting is greatly increased to resemble the shape of the bell.
9. Tie the working stem to its neighbor. Dry in full sun. When completely dry, trim the ends.
10. Add a colorful ribbon to the handle.

Harvest Heart

Hanging Basket

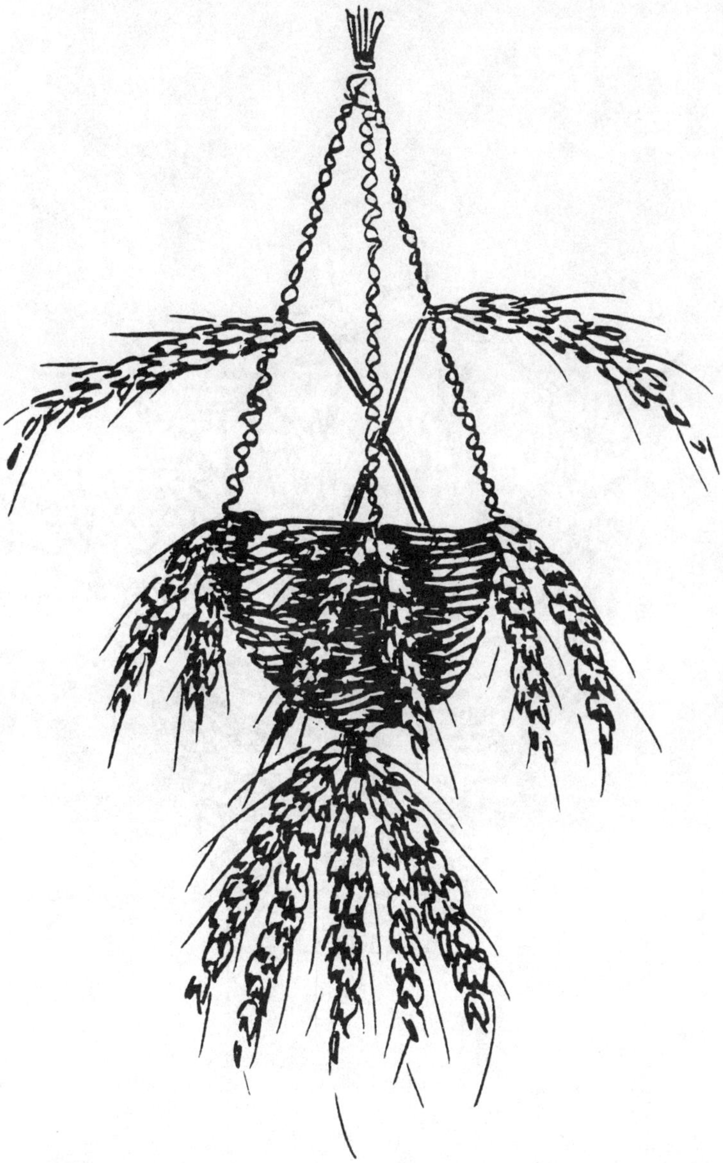

Basket And Wheat

Rope Knot

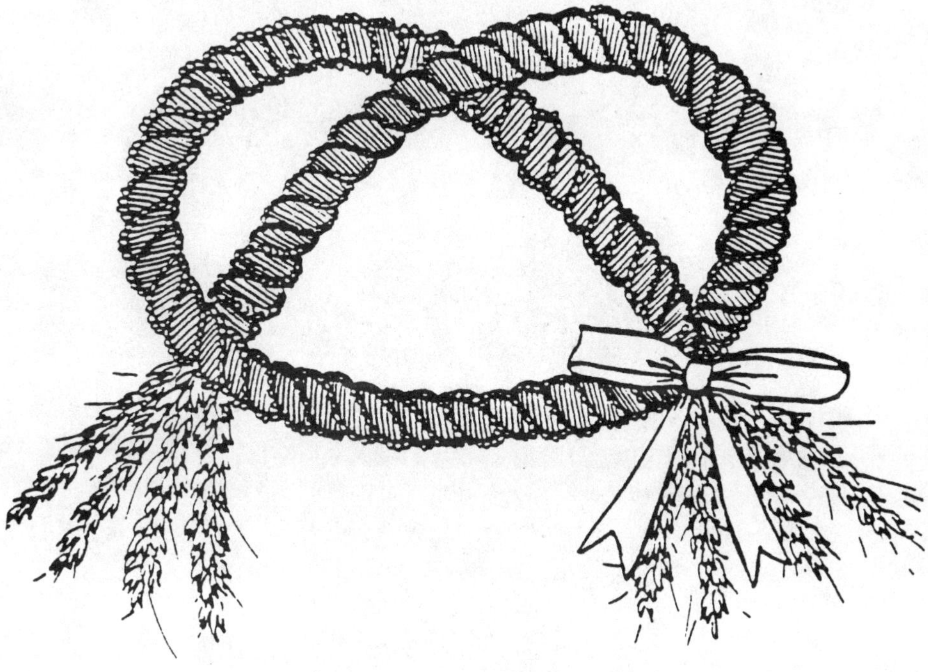

Animals Of Wheat

CHAPTER 9

Horse

Basic Instructions For Wheat Animals

This is an easy beginning Wheat Dolly for children to make.

1. Prepare the wheat by soaking, see page 17.
2. Select 40 wheat straws without heads. The wheat straws should be about 12 inches long with all the necks together at one end. Tie in two places, dividing the wheat stems into three equal parts.
3. The ends of the wheat form the back legs. Bend seventeen wheat stems downward and tie wheat to form the other back leg. The remaining six wheat straws make the tail; however, wait to do this.
4. The front legs are made by bending thirteen of the wheat straws down on each side and tieing at the bottom.
5. The remaining fourteen wheat straws will form the animal's neck and head. Bend the wheat straws to an upward position to form the neck. The head is bent in a downward position. Tie the neck and head securely.
6. Two short wheat straws form the antlers and ears.
7. Raffia, or colorful string is used to hold the wheat straw in place. Begin at the head and neck area, cris-crossing, creating a muzzle effect with the string. For the body, a cross gartering begins at the neck of the animals. Leave a few inches of string free at the end for tying later. Tie securely, then in a spiral fashion bind the body, then down one leg, circle bottom of the leg twice and continue up the leg, forming the criss-cross effect. The string is never cut. Continue with the opposite back leg and up the body to the front legs, and wrap in the same manner. Finally tie the string, leaving a few inches for future use.

(continued)

Basic Instructions For Wheat Animals

(continued from previous page)

8. Trim the feet of the animal and slightly spread out the wheat to allow the Wheat Dolly animal to stand evenly.
9. Several points to remember when designing your wheat animals:
   a. The position of the angle of the head and neck determine the type of animal you are making.
   b. The antlers and ears also tell what kind of animal it is. The position of the ears affect the animals alertness.
   c. Horns may be plaited, then attached in a spiral.
   d. The angle of the nose and the head's length, which may be trimmed, give more indication as to the animal's personality.
   e. The tail's angle, as with that of the neck, give clues about the animal: a curly tail, a straight tail, the tail in and up or a down position.
   f. Add a ribbon and or a bow to finish it off.

A Child's Wheat Straw Toy

Swan

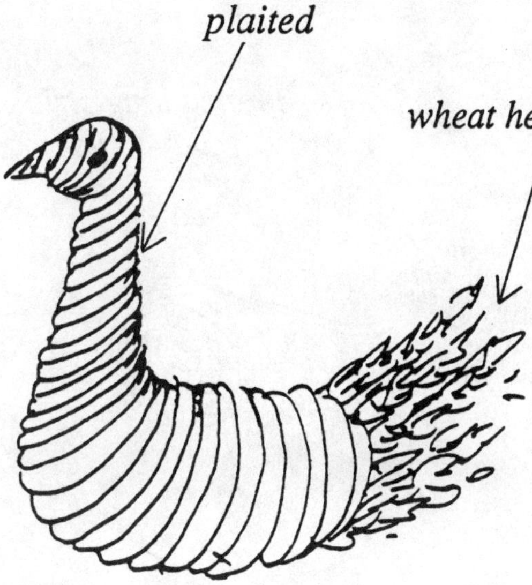

Giraffe

Bird

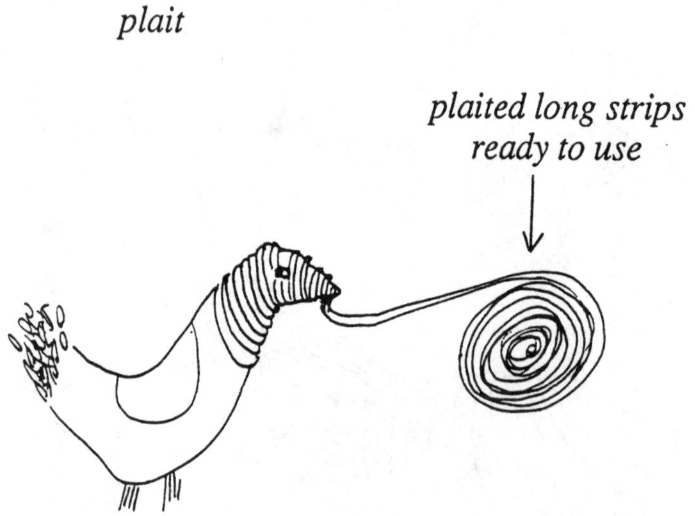

Bull

Sheep

Wheat heads form the body.

Horse

Reindeer

Donkey

Goat

Scandinavian Goat

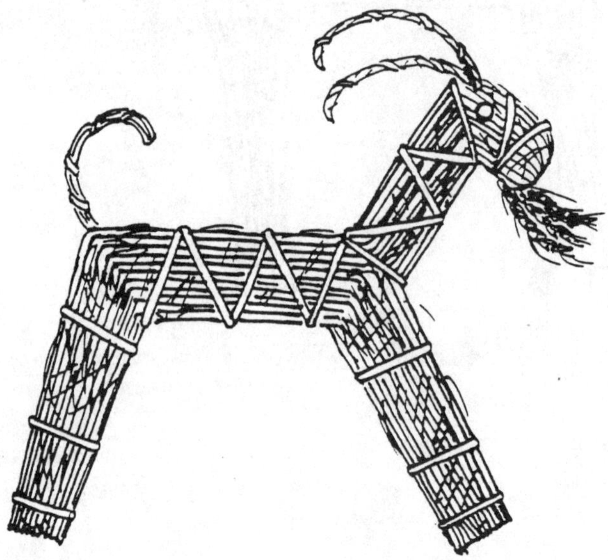

Turkey

Duck

Figures Of Wheat

CHAPTER 10

Wheat Lass

Woven Lady or Spiral Lady

Woven Lady or Spiral Lady

This Wheat Dolly is an excellent example of "increasing to widen" and "decreasing to narrow," a technique which gives the Wheat Dolly its own special personality.

1. Prepare the wheat by soaking. See instructions on page 17.
2. Tie nine wheat straws at the neck. Several may have wheat heads for the lady's hair.
3. Plait a small spiral for the head. increasing then decreasing the plait.
4. Plait 3 rounds of five straw plaiting, keeping the diameter uniform for the neck area.
5. Increase the plaiting for shoulders and upper portion. Reduce plaiting to make the waist.
6. Continue to increase plaiting until a bell-like skirt is made. Tie in the working wheat straw and cut off all protruding ends.
7. The arms are made by carefully pushing fine wheat straws without heads into the body at the shoulder area. The stem ends extend through the body and out the other side about one half inch. On the right side, another five wheat straws are attached to the left side. Tie both ends with the clove hitch knot. The extended ends on each side will be covered with the plaiting of the opposite arm.
8. The fine straw spiral is used to make the arms, increasing and then decreasing gradually to the wrists.
9. Tie the ends with a clove hitch knot.
10. Extend straws beyond the tie to form the ends.
11. Fold the arms to the front and tie the hands to three wheat heads.

Angel With Flowers

Modern Miss

This Modern Miss stands on a wooden core. Begin at the head with a spiral plait increasing and decreasing for the body shape. Attach plaited arms.

Wheat Maiden

Girl On A Swing

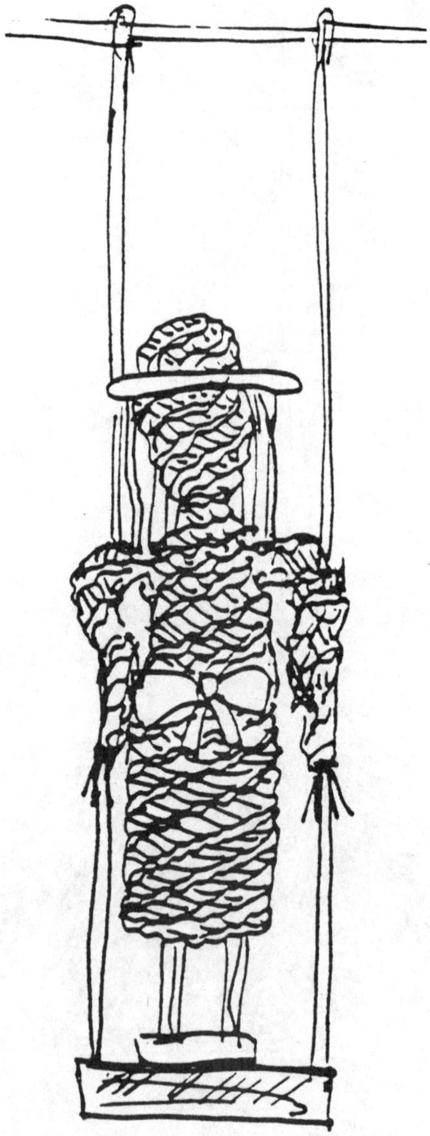

Wheat Lass

Harvest Doll

Straw Angel

Twenty-five pre-soaked wheat stems without heads form the angel's body. The head, arms and halo are plaited. The wings are attached to the waist.

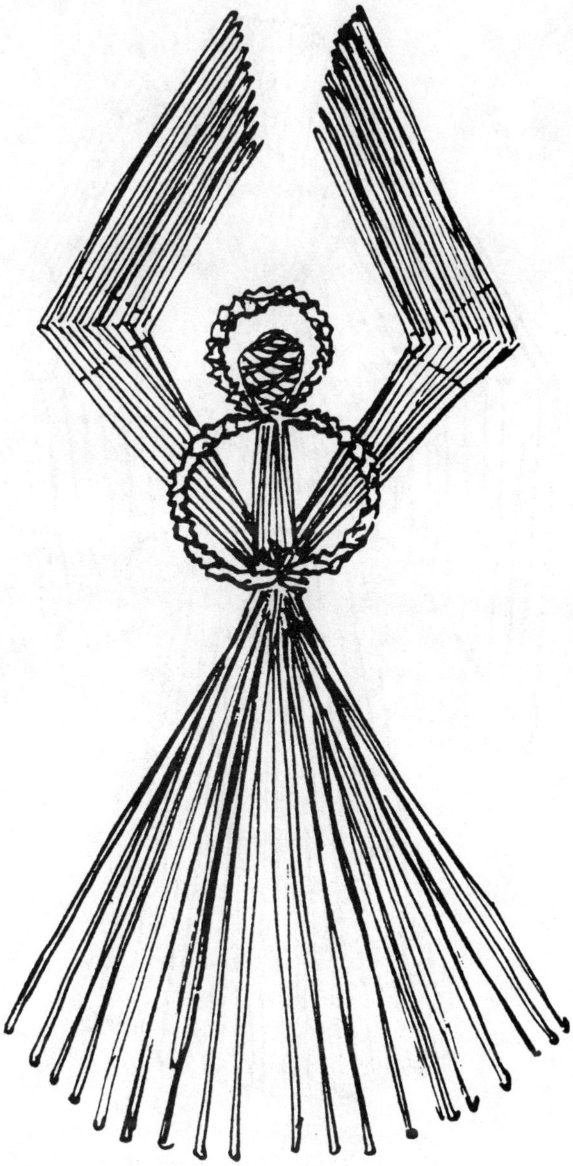

Wheat Angel with Wings Down

Wheat Miss

Tied Straw Dollies

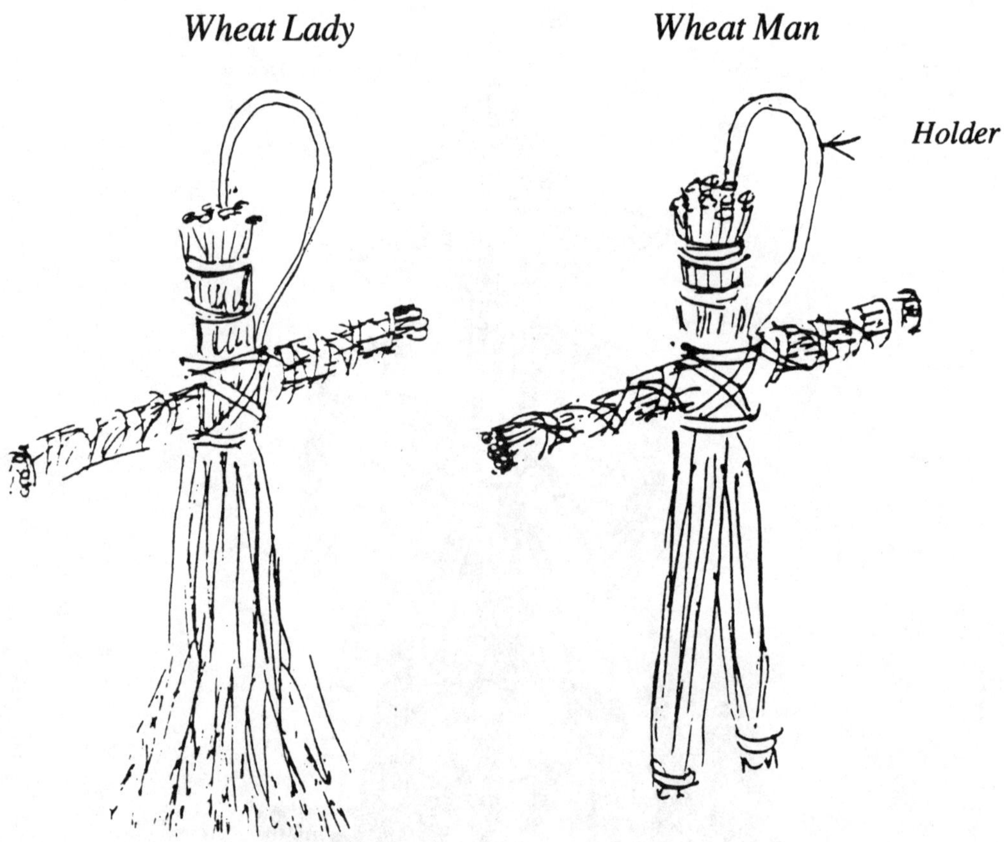

Christmas In Wheat

CHAPTER 11

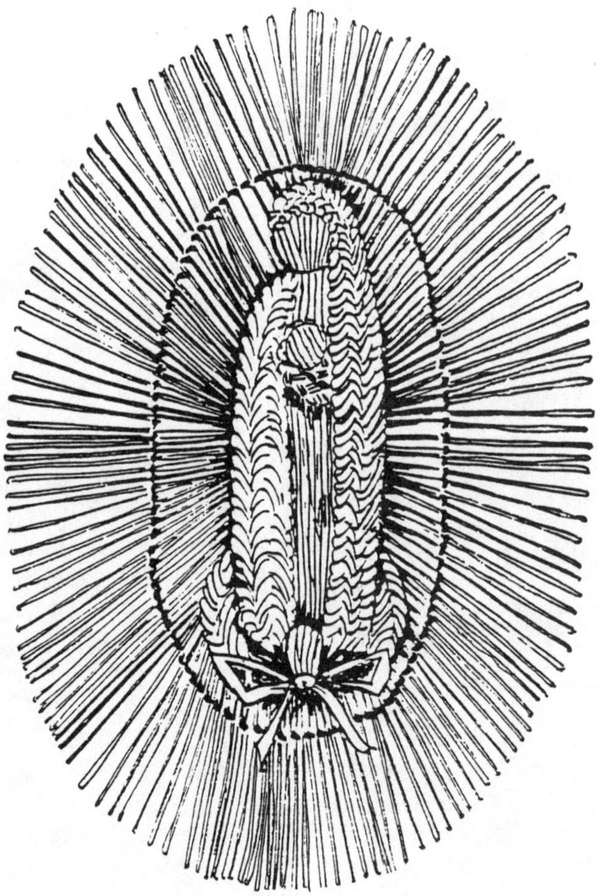

Mother and Child

Tree Decorations

Assorted Wheat Dollies make charming Christmas Tree decorations.

Christmas Tree

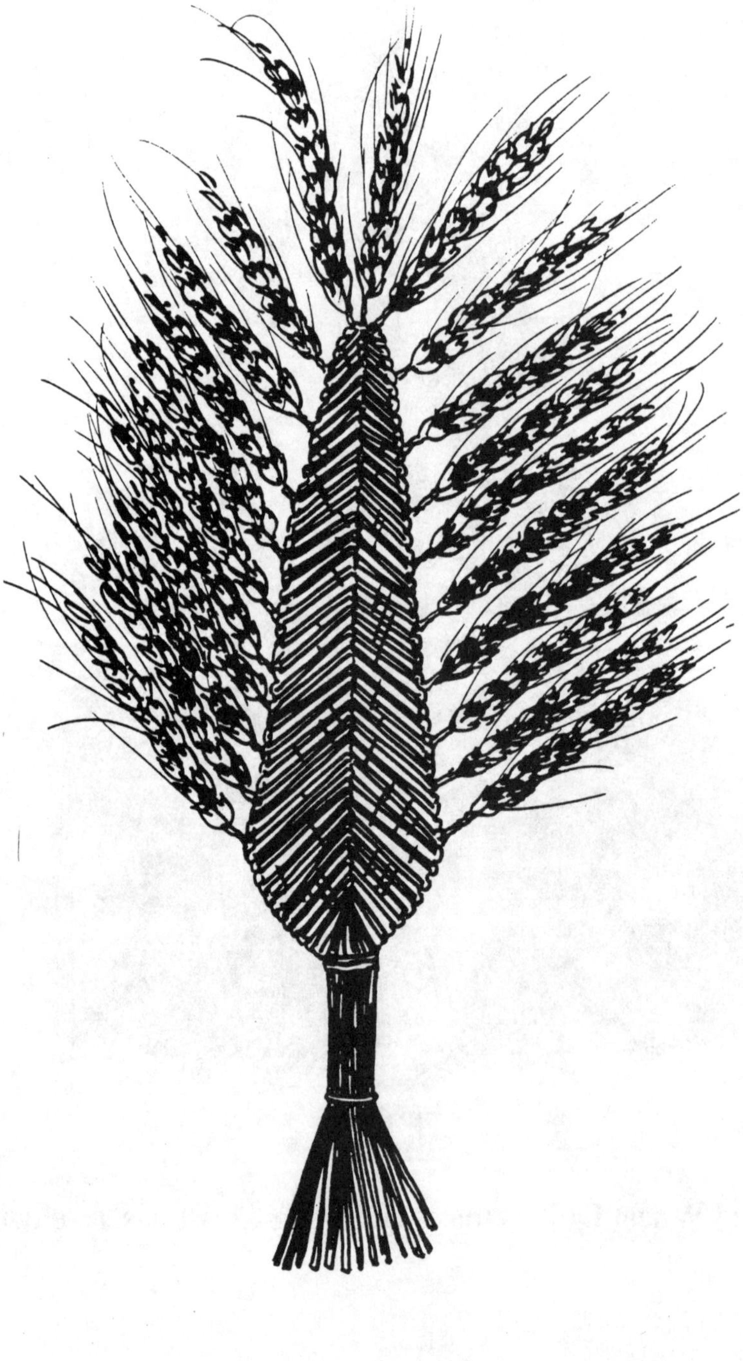

Christmas In Wheat

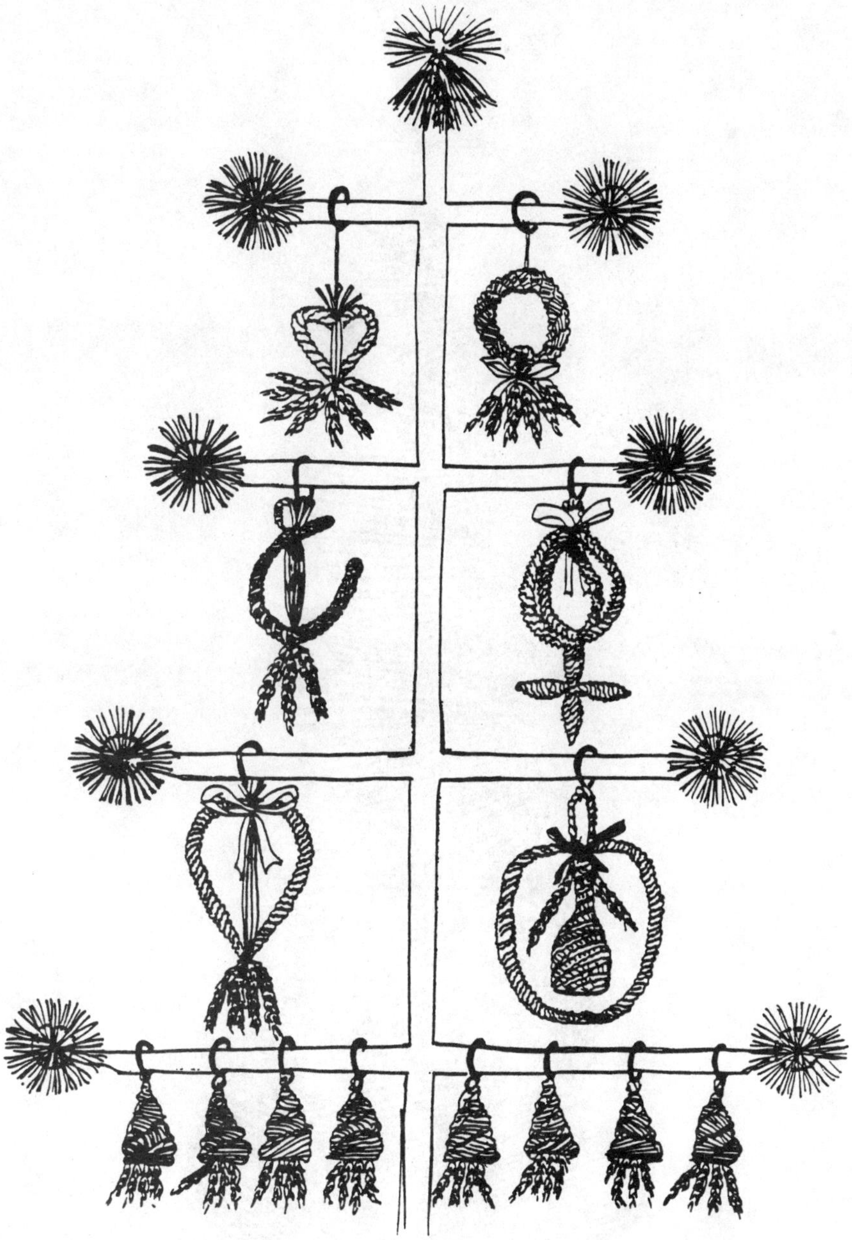

Round Christmas Tree

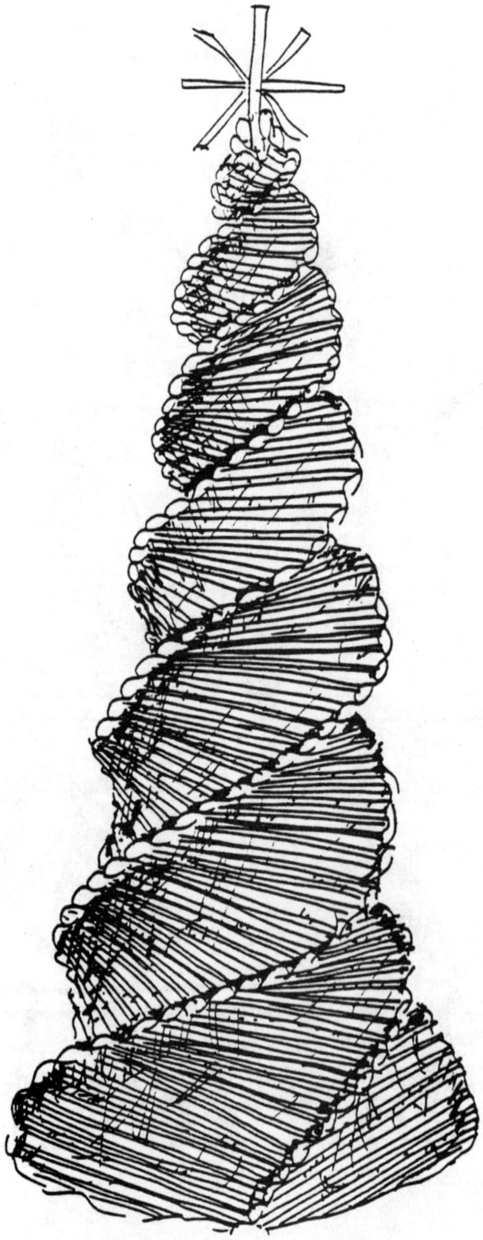

Christmas Candle

Birch Tree

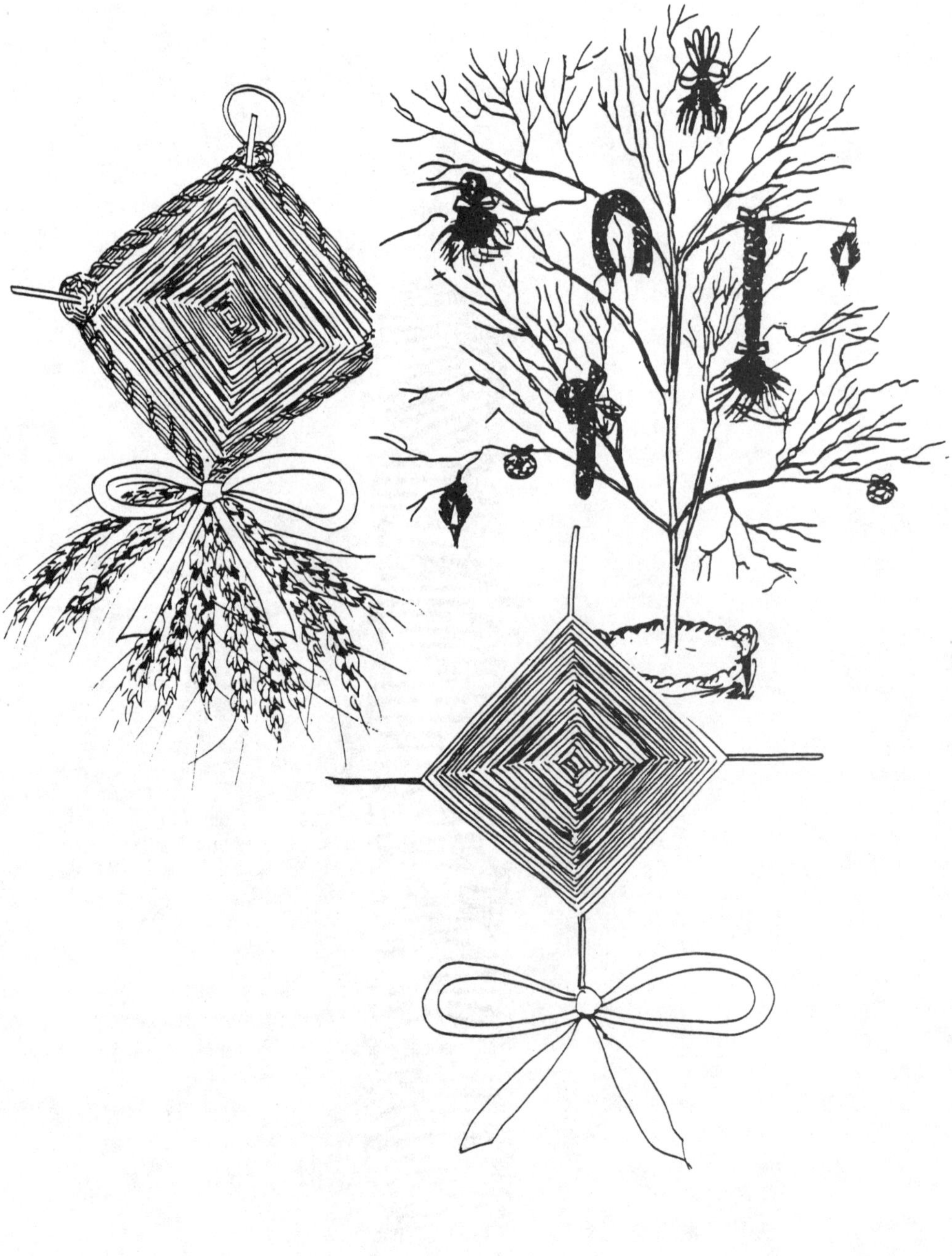

Eight Pointed Star Variation

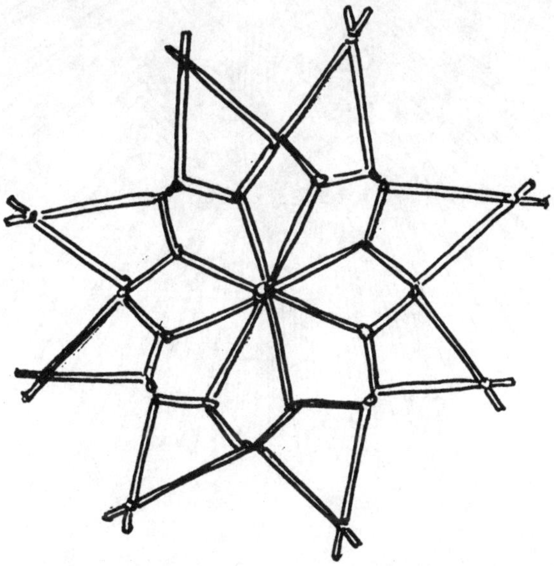

Mother And Child

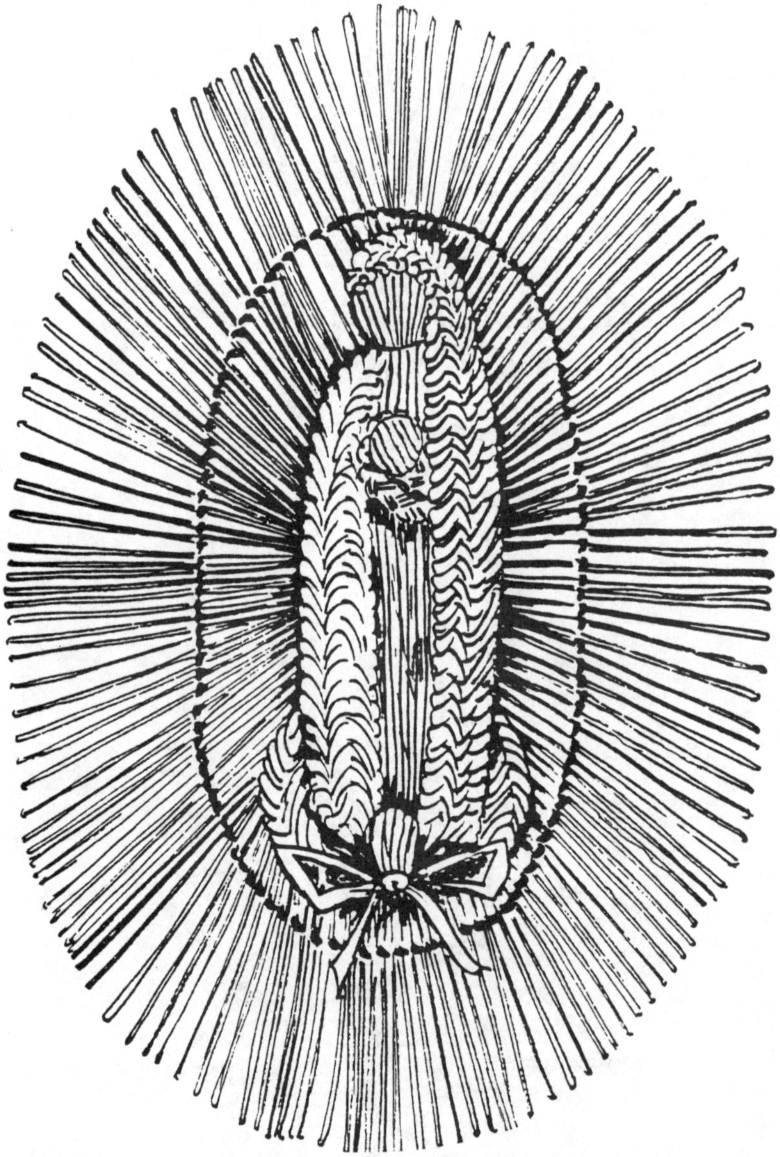

Petit Wheat Angels

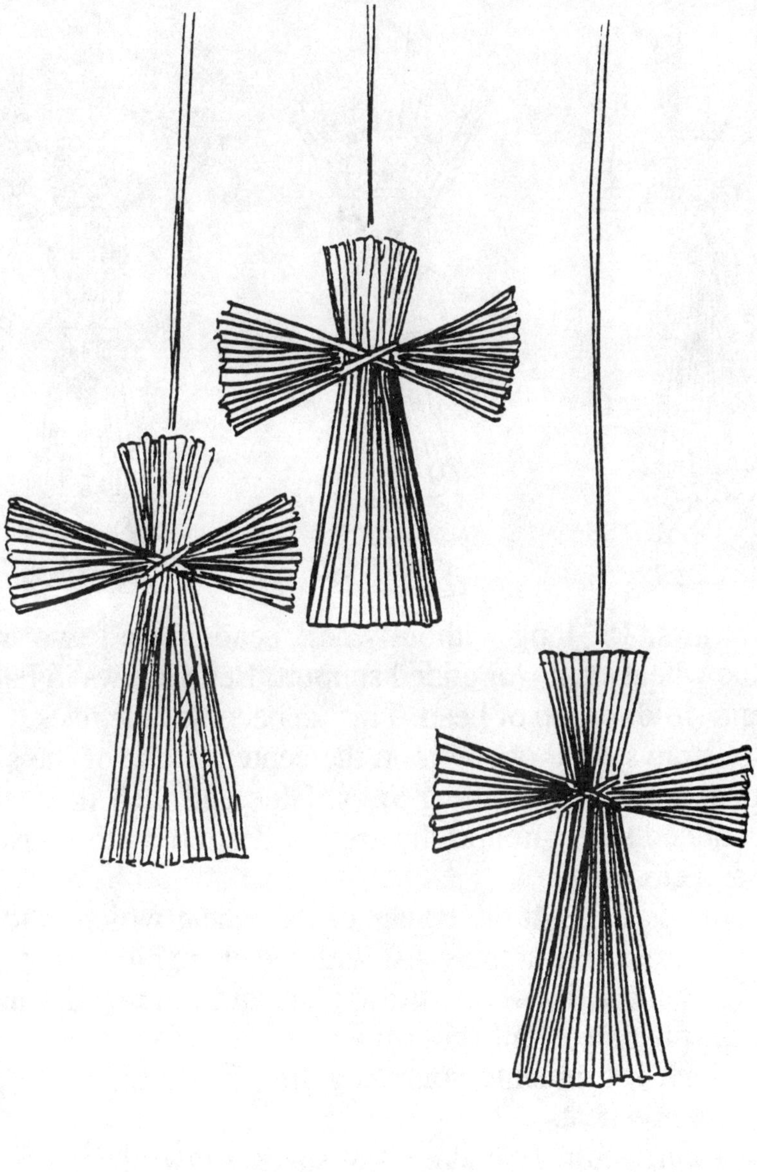

Angelic Musicians

Angelic Trumpeter *Praying Angel* *Reverent Angel*

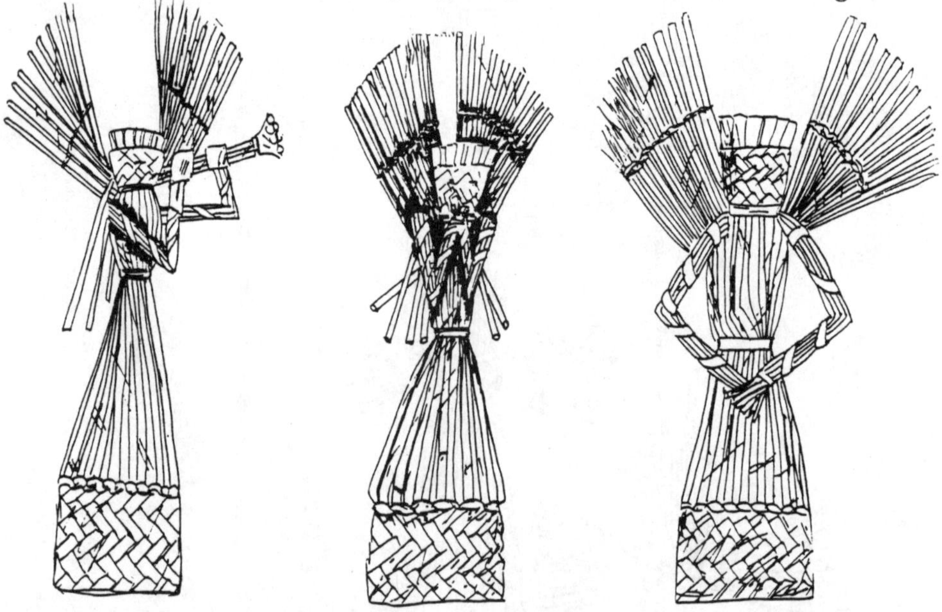

1. Use 30 straws, 15" long without wheat heads. Insert two wires into two of the wheat stems for added support. Bend straws in half. Tie 2" from bend (fold) or top of head. This tie becomes the neck.
2. Select six straws. Add one wire in the center of one of the six straws for arm support when bending arms. Glue and then tie at the wrist. Wrap an added straw around the arms if desired – otherwise, just tie at wrist and elbow.
3. Insert the arms through the center of the wheat which is to become the body. Tie securely at the waist with the clove hitch knot.
4. Make face with a flat seven straw plait and glue to the head area. Ends of the plait are evenly cut off.
5. Decide which instruments your Angelic Musicians are to play and design them in wheat.
6. For added trim you may make a flat seven straw plait to be glued at the bottom of your musician's robes.
7. The wings are made the same as for the angels.

Christmas In Wheat

Basic Angelic Angel

Bass

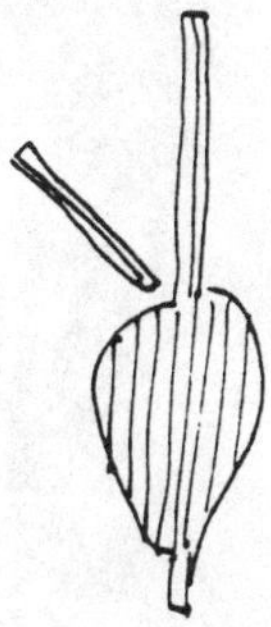

Drum

Cymbal

Guitar

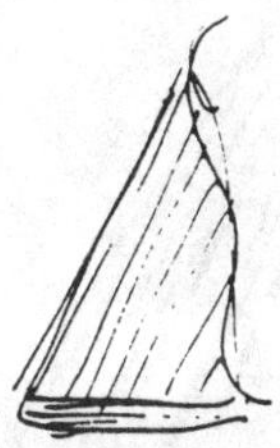

Harp

Sax

Trumpet

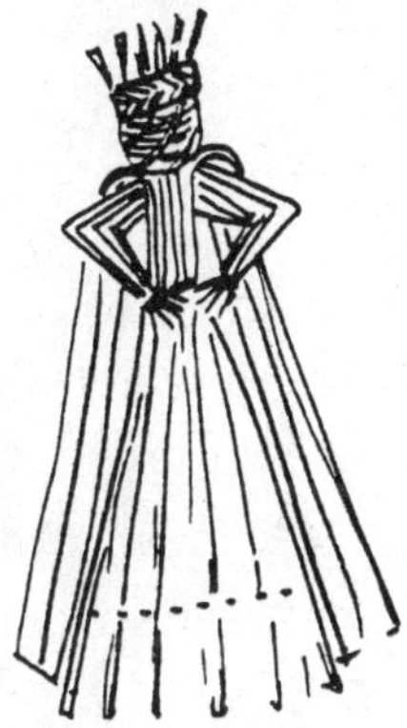

Three Wise Men

Creche Variation

Creche Variation

Angel With Flowers

Musical Wheat

CHAPTER 12

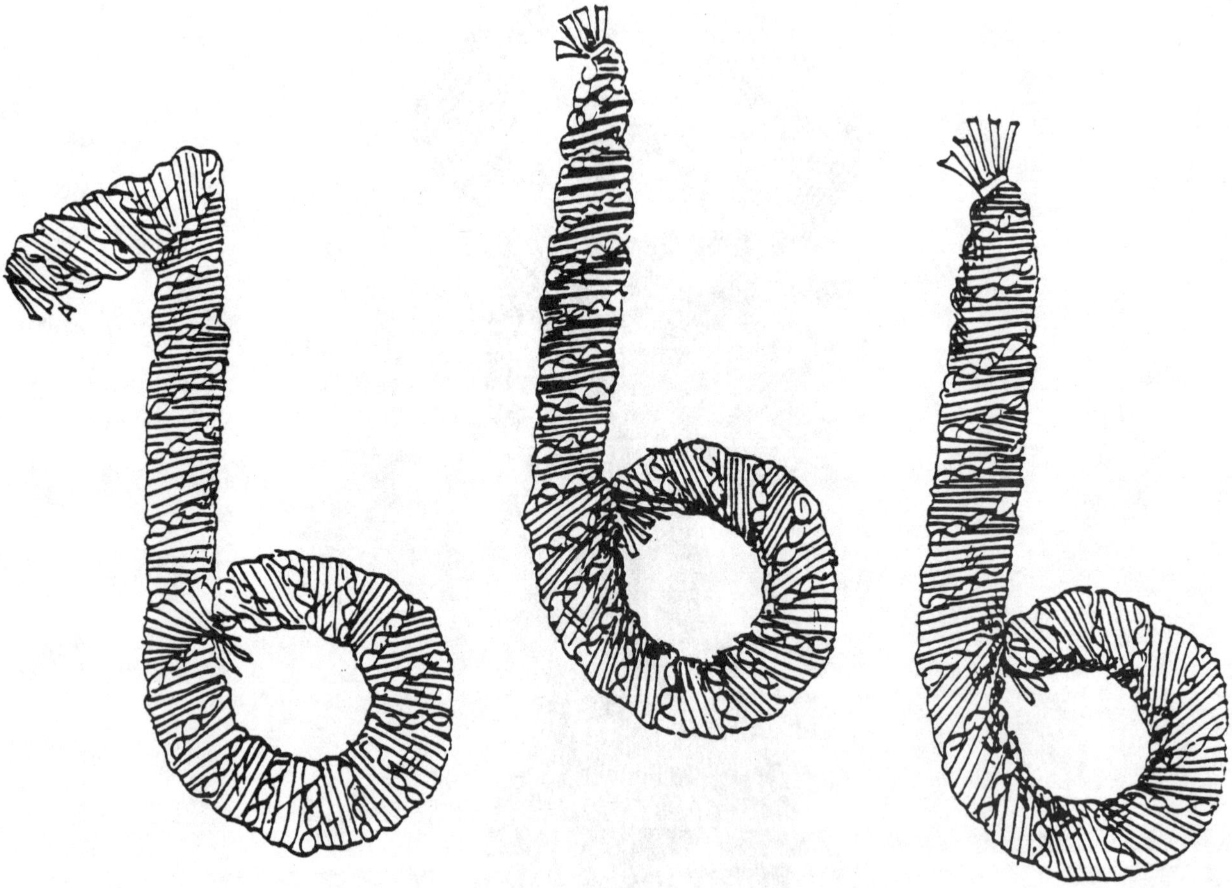

Musical Notes

Golden Clef

The clef is constructed with the five straw plait over a core. While the work is still damp, it is shaped into a clef.

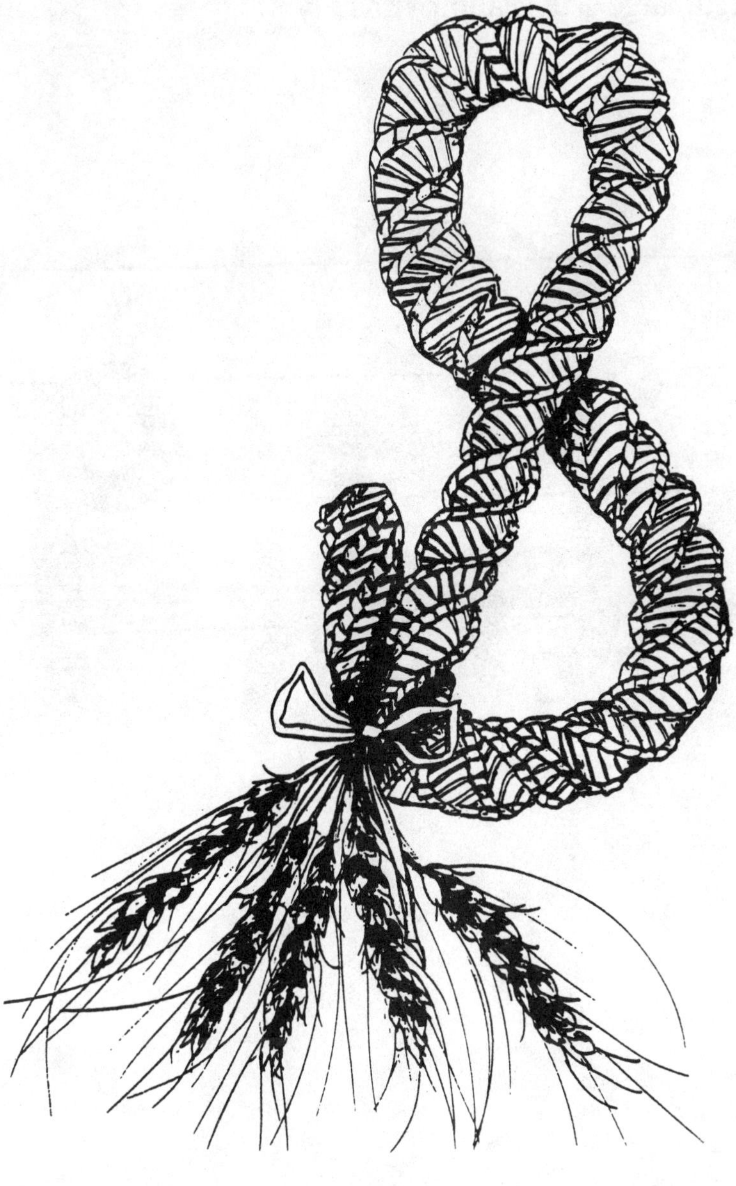

Musical Notes

Musical Notes are made in proportion to the Golden Clef. An entire line of music can be produced for some special occasion, or single notes can be given. They are suitable as gifts for choir members, musicians, or anyone enthralled with any kind of music. Small size is usually about 4" high; large size is approximately 8".

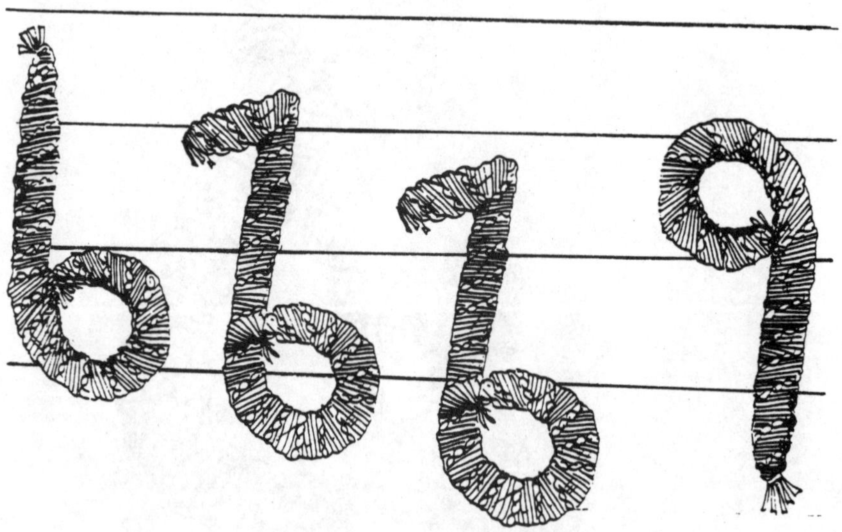

Musical Notes

Plaited Numbers

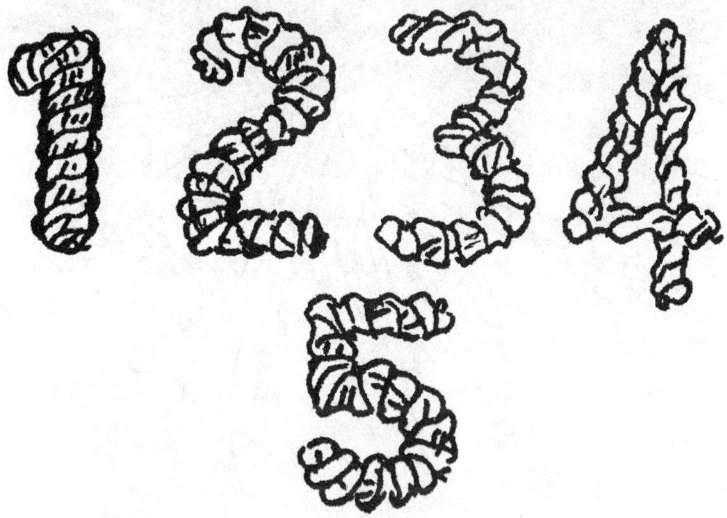

Wheat Wreaths

CHAPTER 13

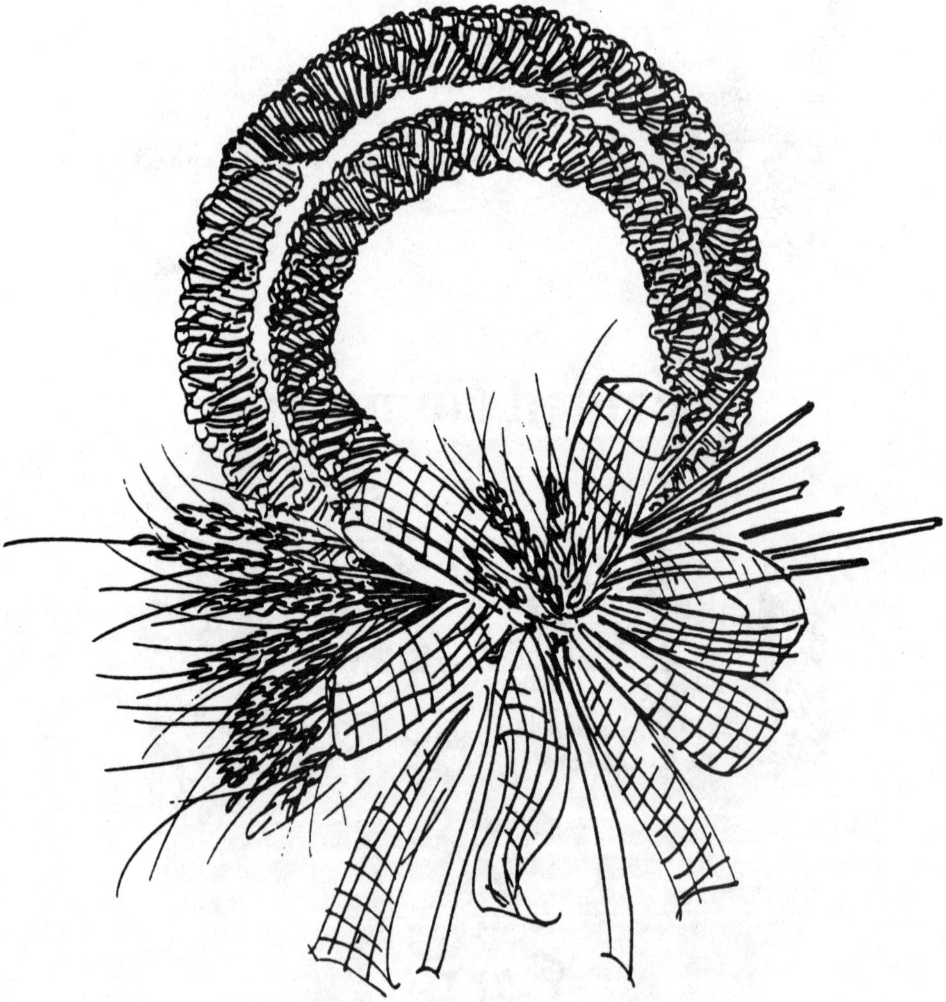

Single Woven Wreath

(Classic Wreath)

Photograph Of Wreath

Circle-Of-Life Wreath

The wreath has special meaning – continuation of life and life everlasting. The wreaths themselves are long lasting with proper care. They can be hung on the outside or inside of doors; over doorways or on walls in place of pictures; on cupboard doors; over the arm of a wooden chair, or they can simply be placed casually on a mantel or laid on a coffee table.

The 36" Circle-of-Life Wreath is a cylindrical ring of bound wheat straw. The frame for it is an old tire tube cut in half.

Petit Wheat Wreath

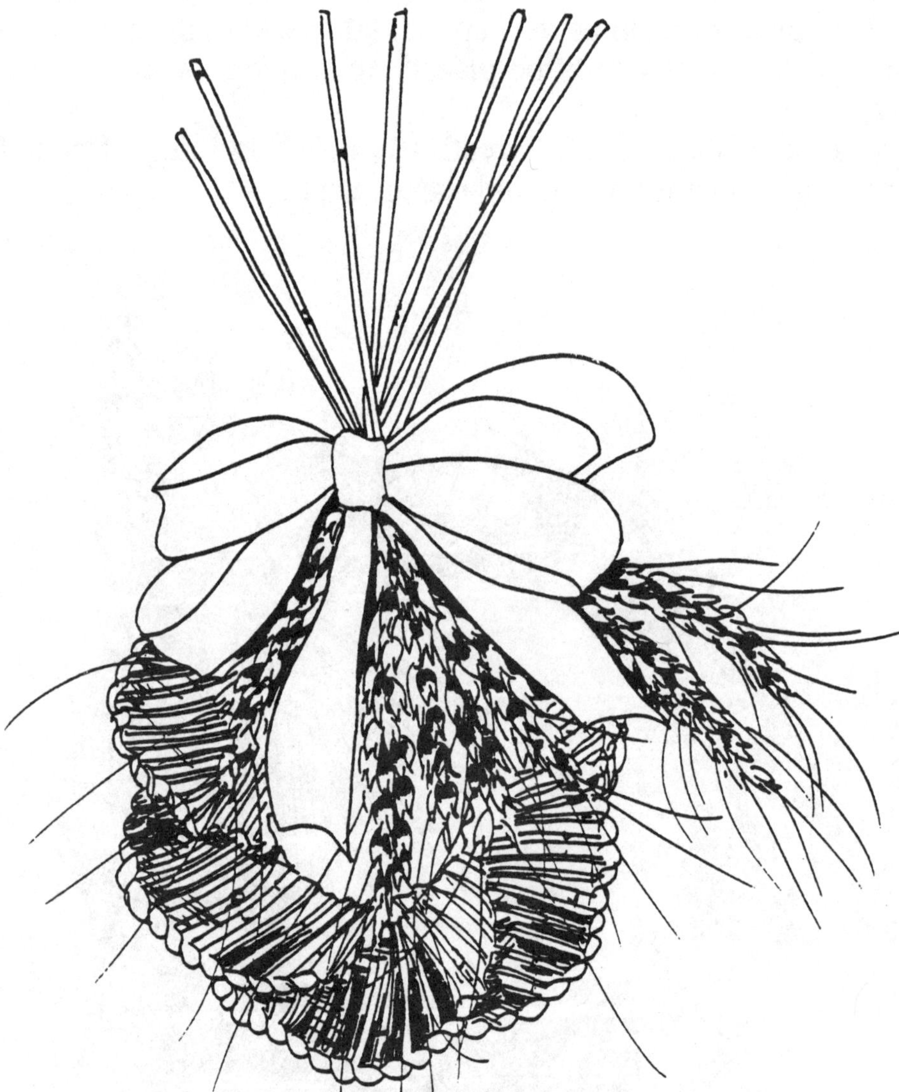

Pheasant and Straw Wreath

Straw wreath enhanced by beautiful pheasant feathers, wheat stems and heads, and ribbon.

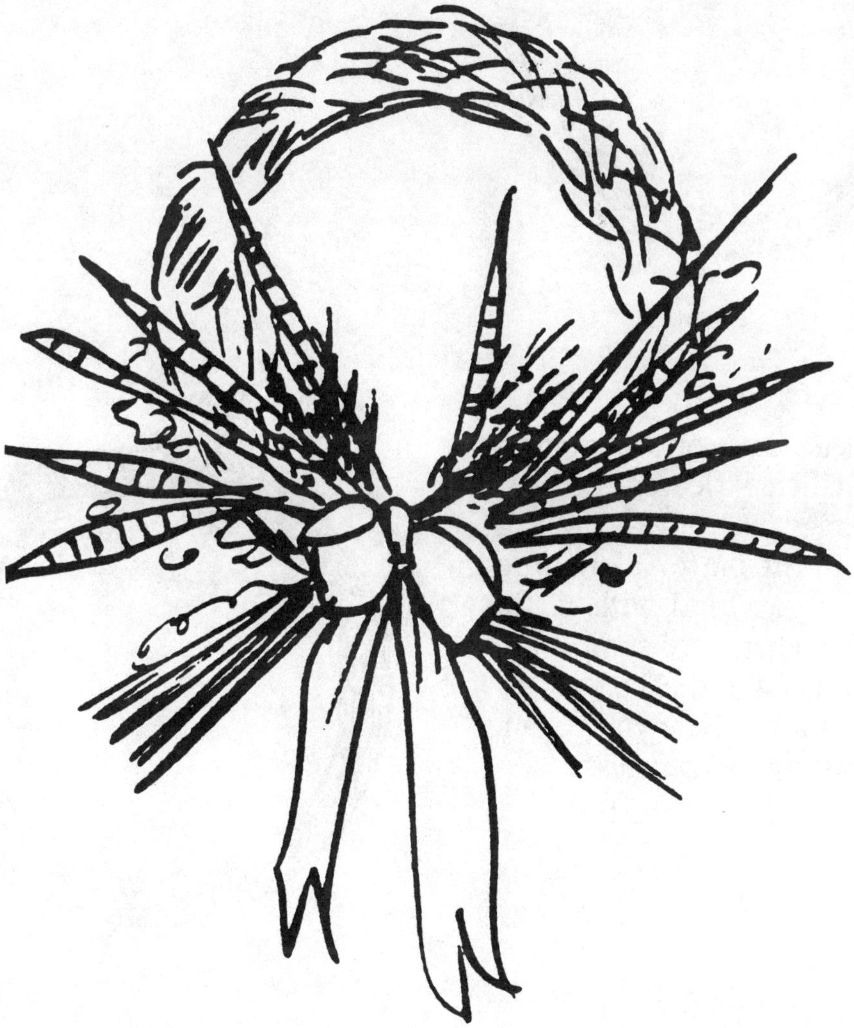

Christmas Wheat Ring

1. Pad a metal ring of any size with pre-soaked straw that is easy to handle. Tie wheat to ring with natural colored string or raffia.

2. Wheat, oats, or a combination of dried flowers, seeds, and leaves may now be added. Hold plant material in one hand as you bind with raffia with the other, until completely covered. Ease the last set of material into where you began, re-adjusting for evenness.

Narrow Woven Wreath

Scandinavian Wreath

Single Woven Wreath

(Small and Large – Classic Wreath)

The small Single Woven Wreath is 6", a couple of inches smaller than the large one. It is ideal for smaller wall spaces, cupboard doors, door knobs. Most dwellings and decorating schemes offer additional opportunities for ornamentation.

Double Woven Wreath

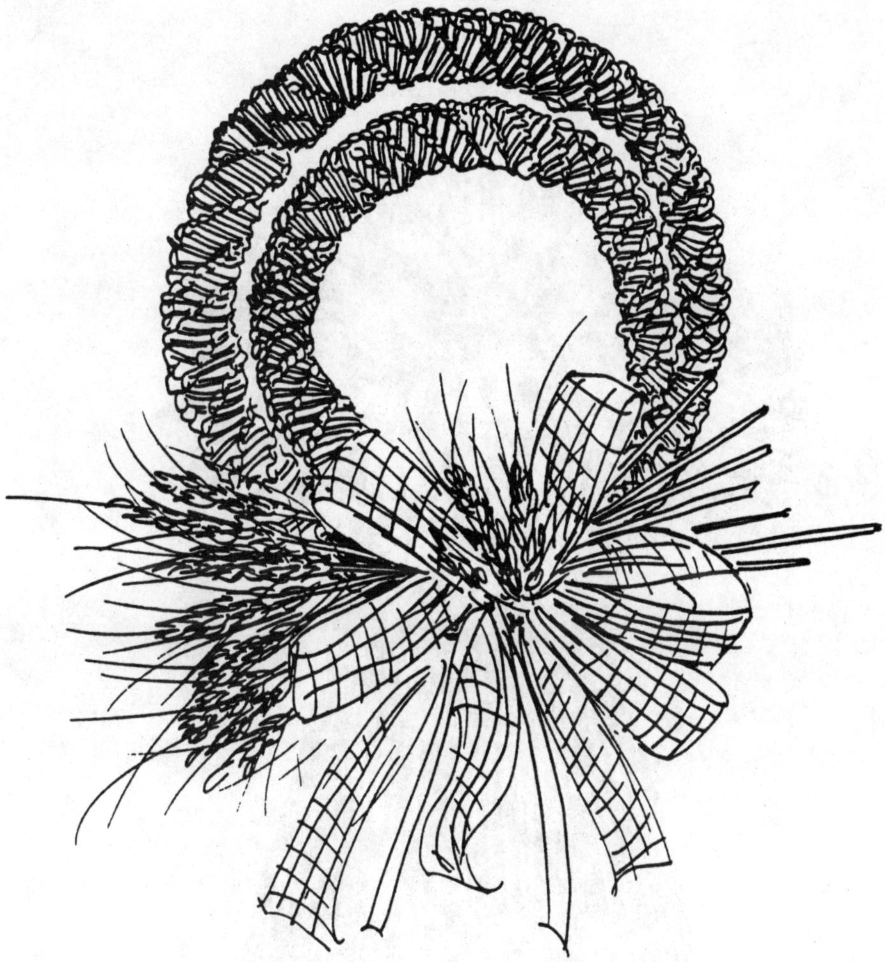

Wheat Bundles

CHAPTER
14

Standing Wheat Sheaf

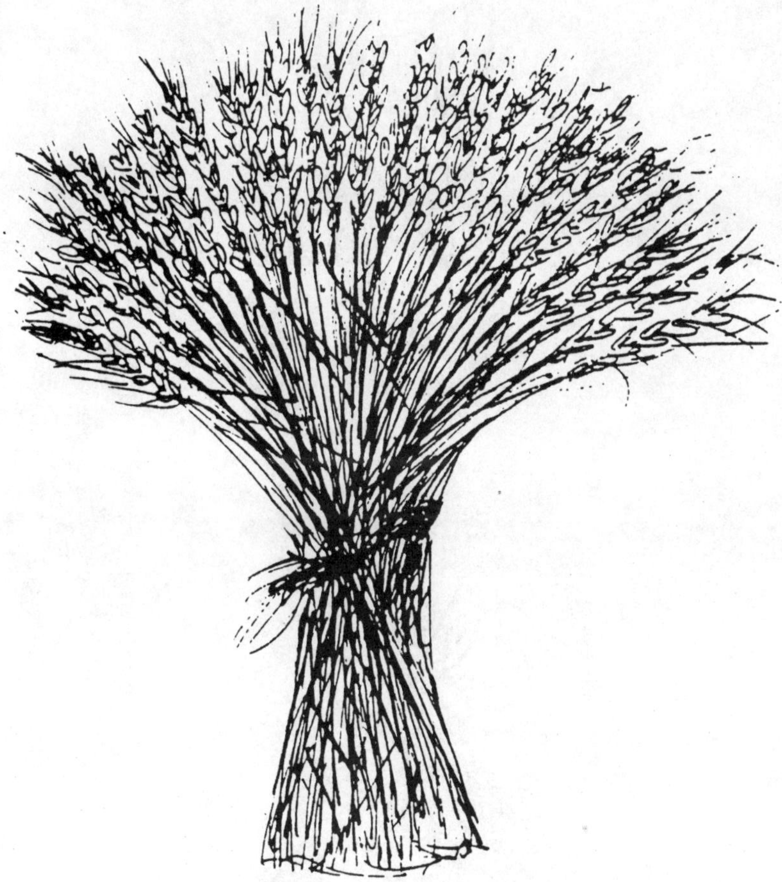

Spring Wheat Bouquet

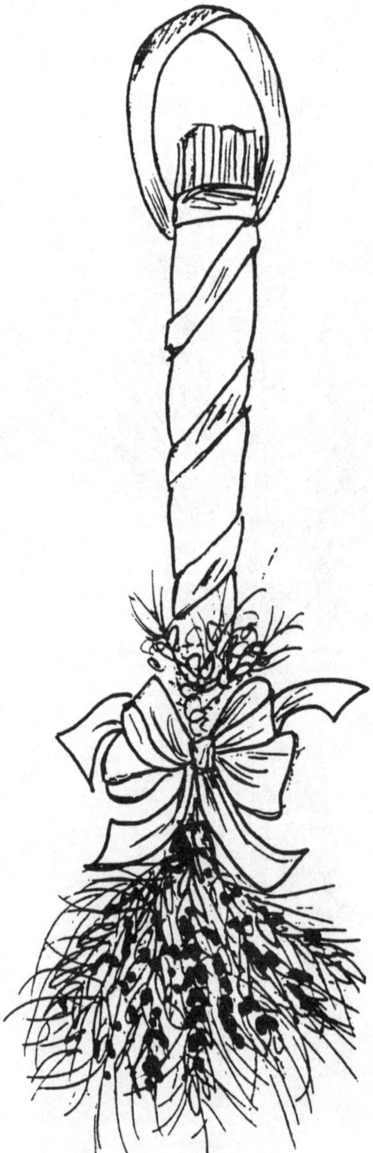

Spring Wheat Bouquet

Use long wheat stems. The size of the wheat bundles can be as small as a quarter pound or as large as one pound. Tie first with strong string, then wrap with wide ribbon. Decorate with bows, flowers birds and nests of your choice.

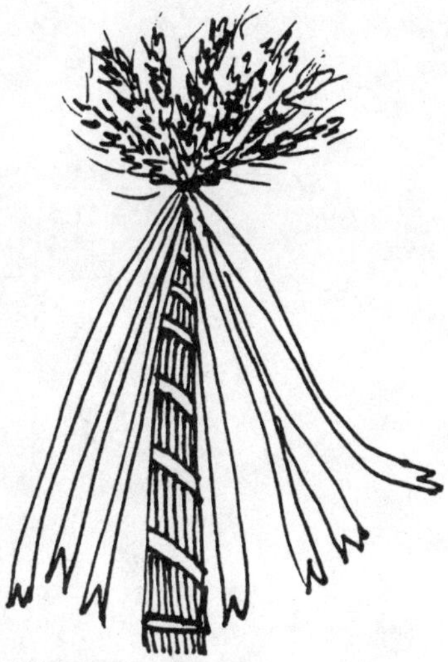

Special Occasion Wheat Dolls

CHAPTER 15

Double Bells

Harvest Cross Variation

Easter

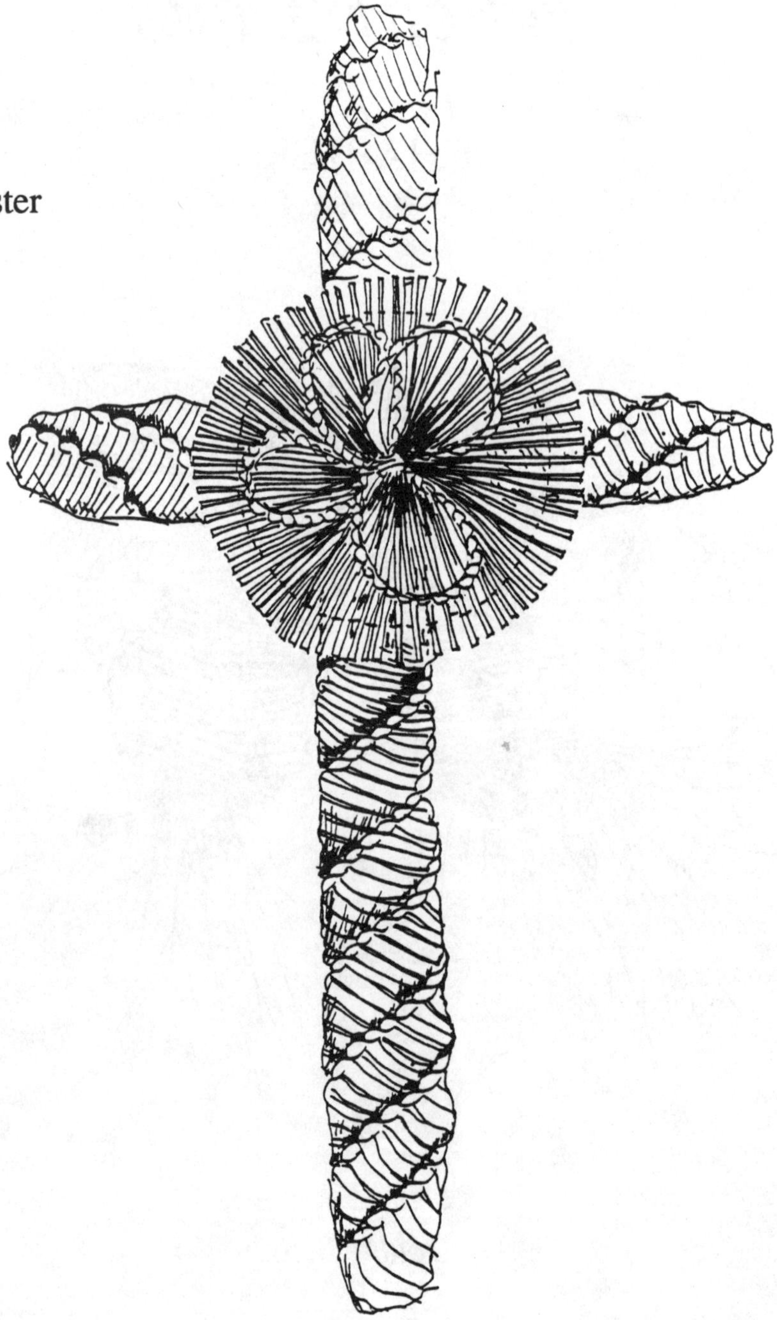

Hearts

Examples of different plaits.

Wedding Cake Top

BIL

PEO

The B.I.L. Heart

1. The heart is plaited with a wire insert for shaping into a heart. When completed this can be glued onto a cardboard or heavy paper cut-out of the same size as the heart.
2. Letters of your choice can be printed on the paper.
3. A floral design could be drawn if letters are not used.

PEO

The star is made with a flexible wire as a core, using the five straw plait or a plait of your choice. Plait approximately 15" long, attach ends securely with fine wire. Wheat should be damp enough to bend and shape into a five pointed star.

Shape wheat into star around nails that have been hammered into a thick piece of wood. Join ends with wire and for added strength use plastic glue and glue gun.

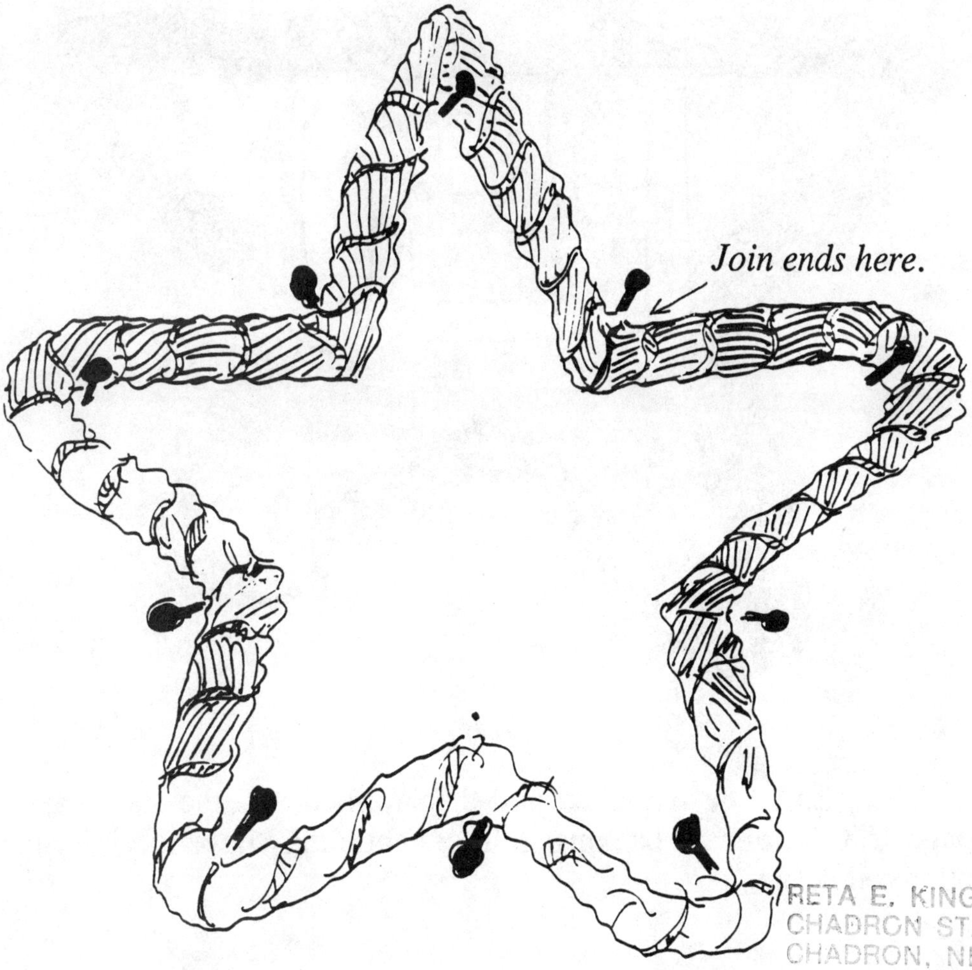

The Star And Its Message

The P.E.O. Star

The P.E.O. Star is plaited and then formed into the star shape, see page 158. Cardboard backing, felt, or fabric may have the information on it you desire.

Decorating The Home With Wheat Dollies

CHAPTER 16

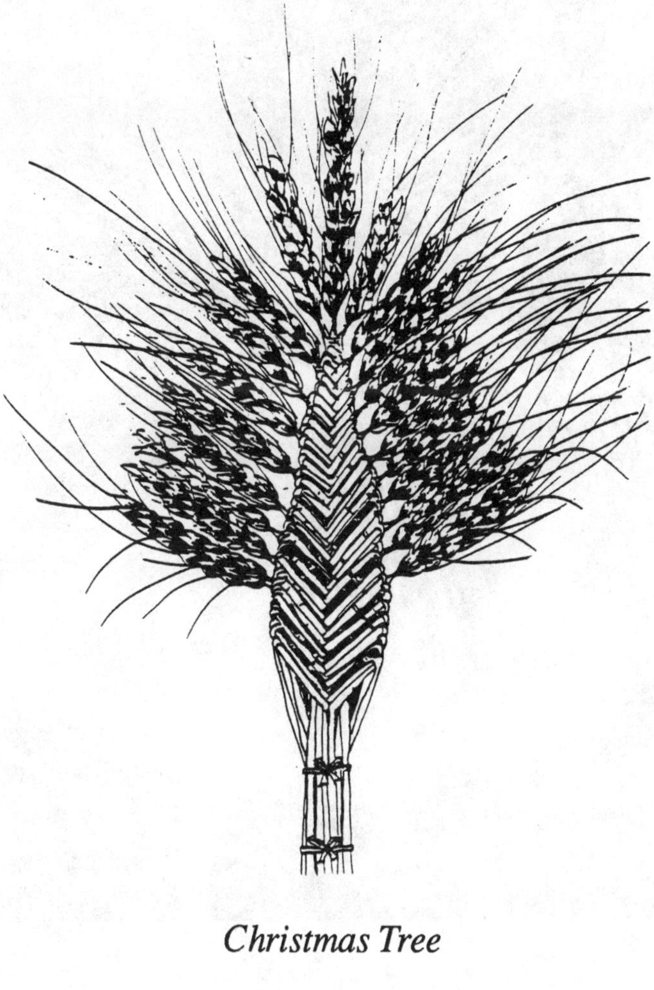

Christmas Tree

Frame Your Wheat Dolly

The Wheat Dolly Mirror

An oval beveled mirror is framed with rows of plaited Gems. A plaited rope outlines the mirror.

Compositions In Wheat

CHAPTER 17

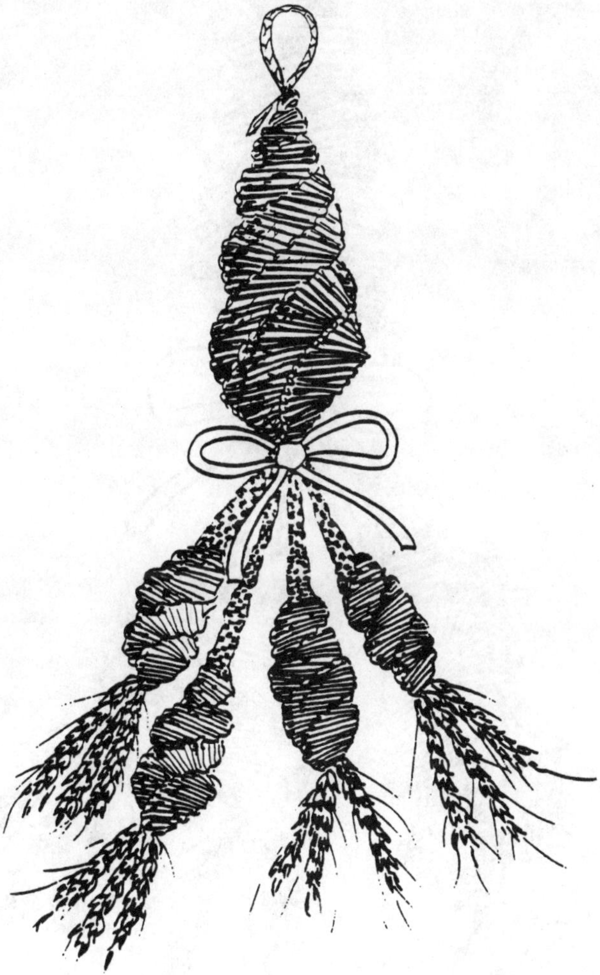

Wheat Spirals with Tassels

Wheat Spirals With Tassels

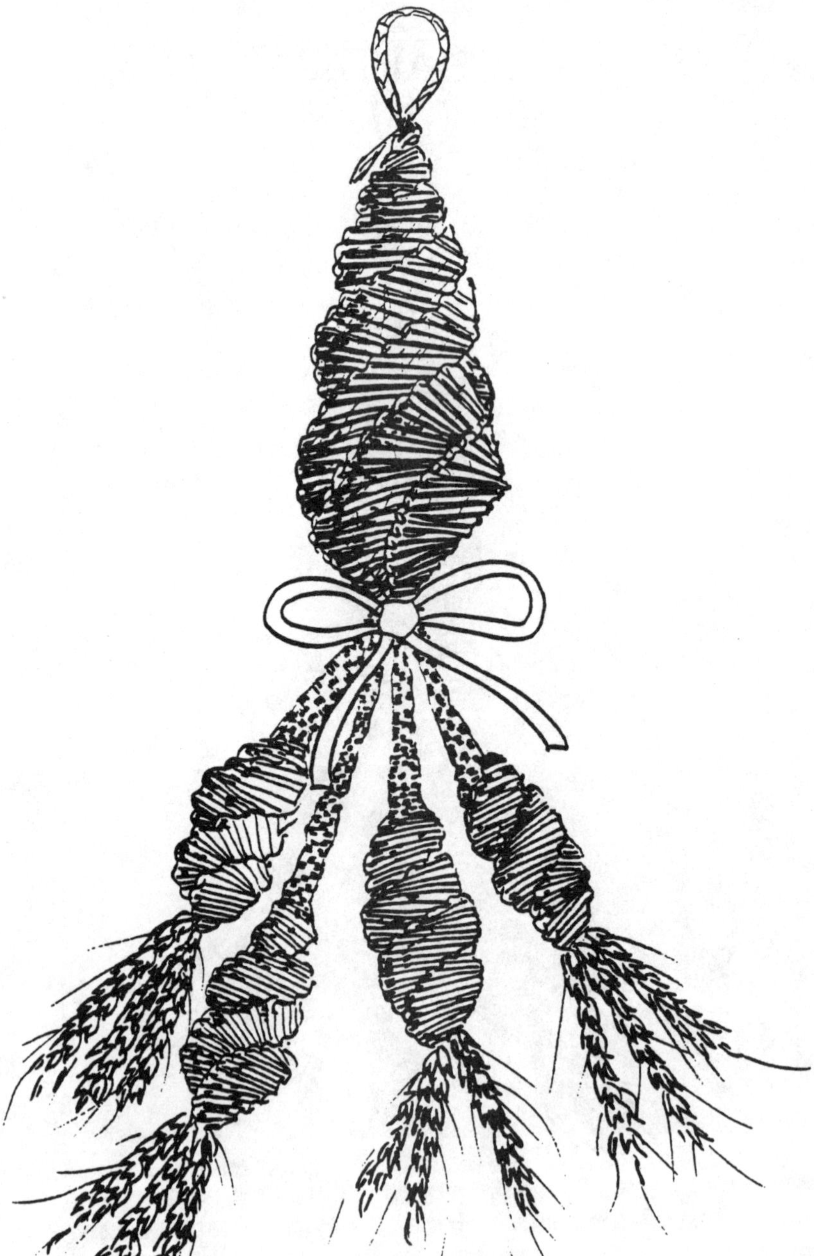

Wheat Sheaf And Wreath

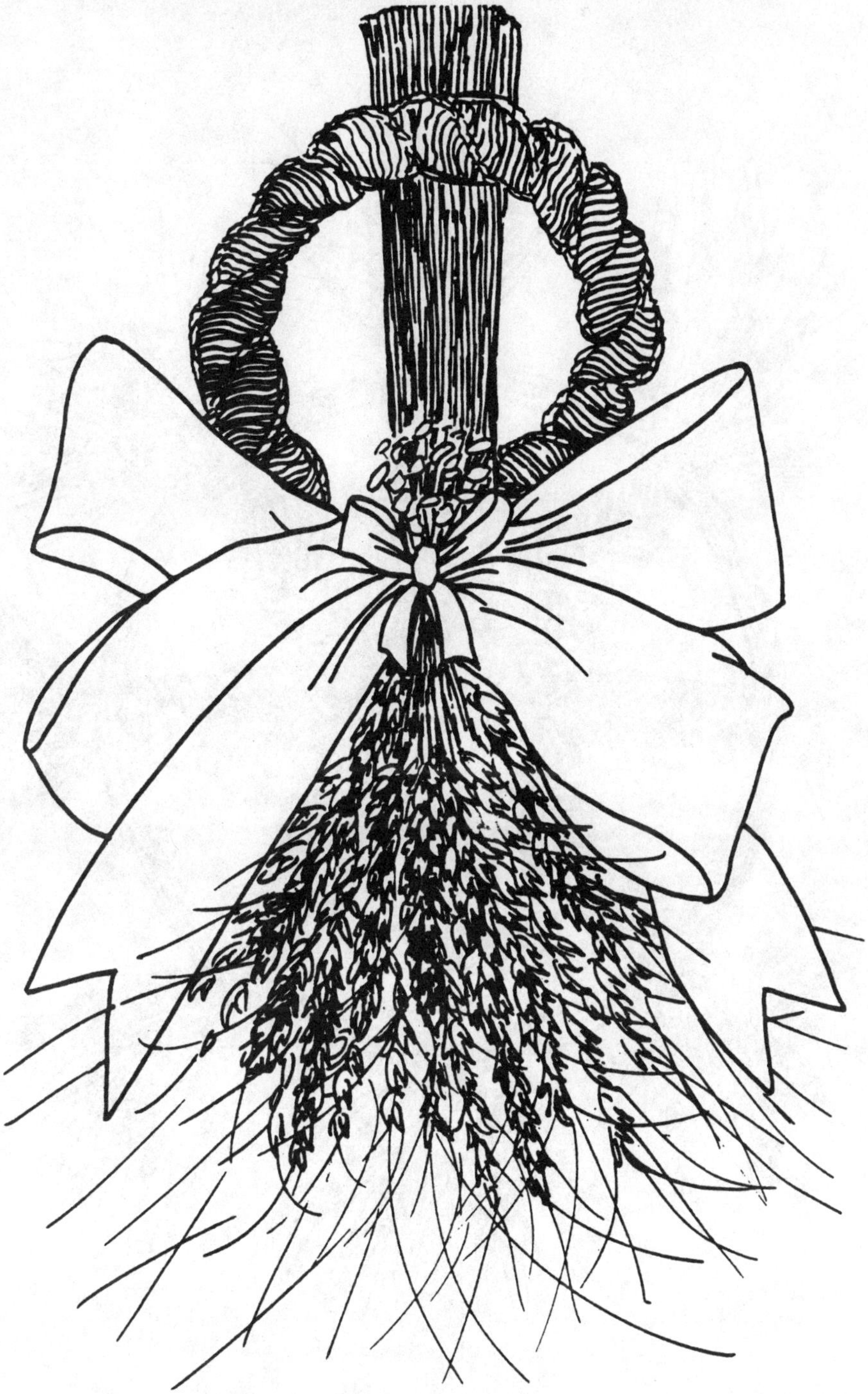

Gems And Christmas Wheat Ring

Petit Sheaves And Hearts

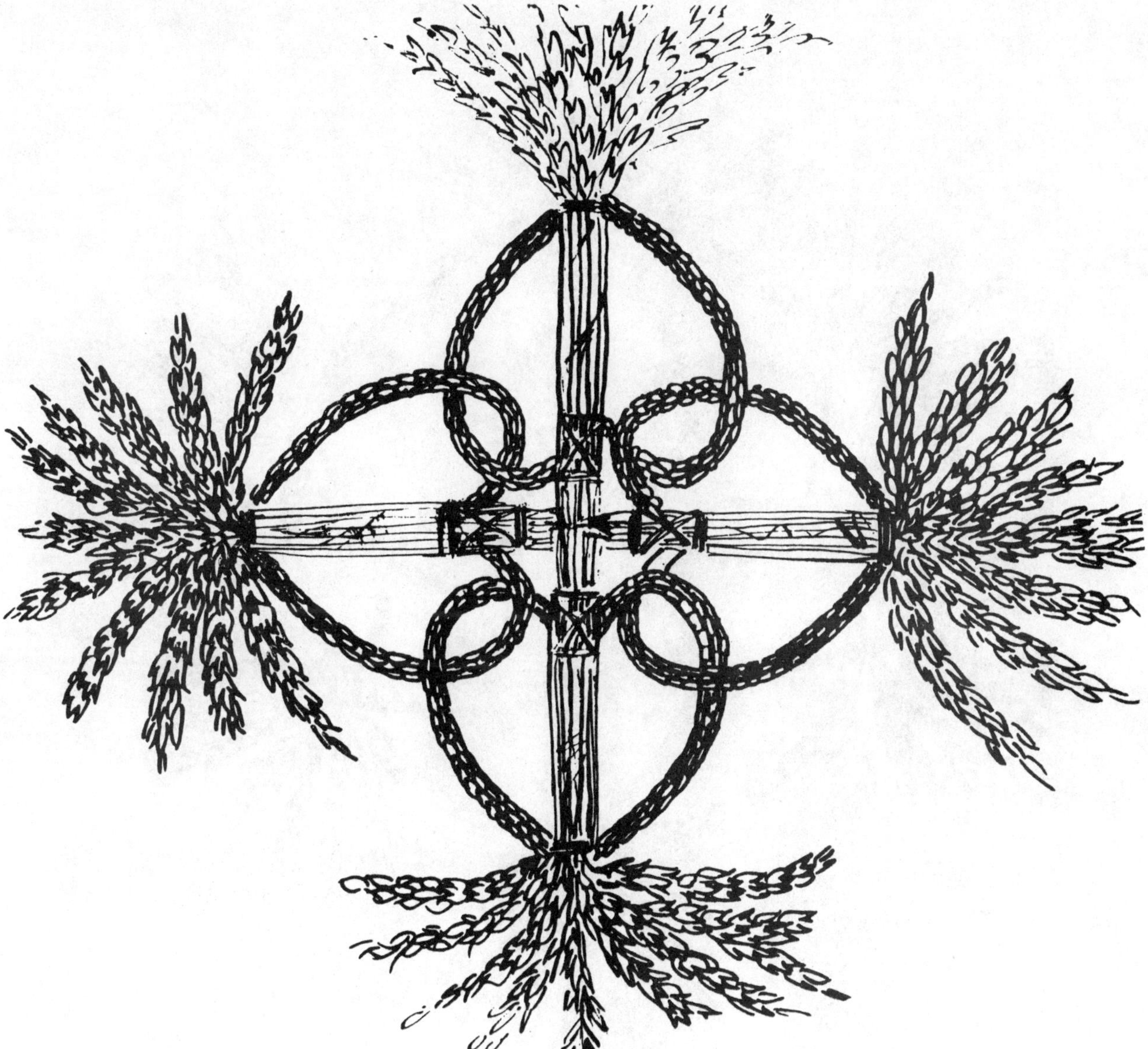

Wheel of Hearts

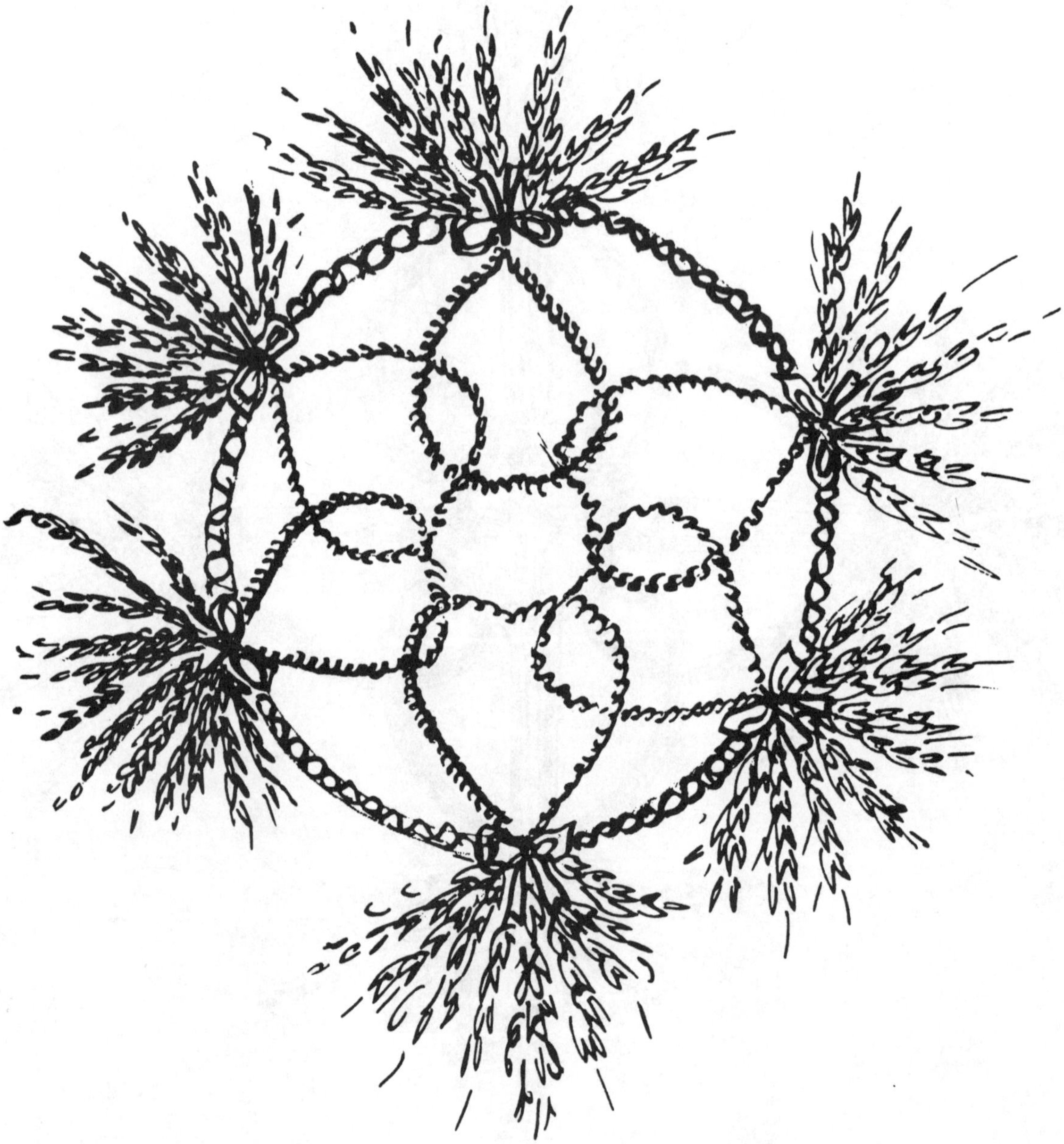

Wheel of Hearts

1. The circle is plaited over a quarter inch thickness of rope. The ends of the rope then are attached.
2. Six plaited hearts are secured to the main circle. The upper part of each of the six hearts is intertwined. The ends are then attached and tied at the bottom part of the heart to the main circle with a ribbon.

Wheat Chandelier

Lazy C Composition

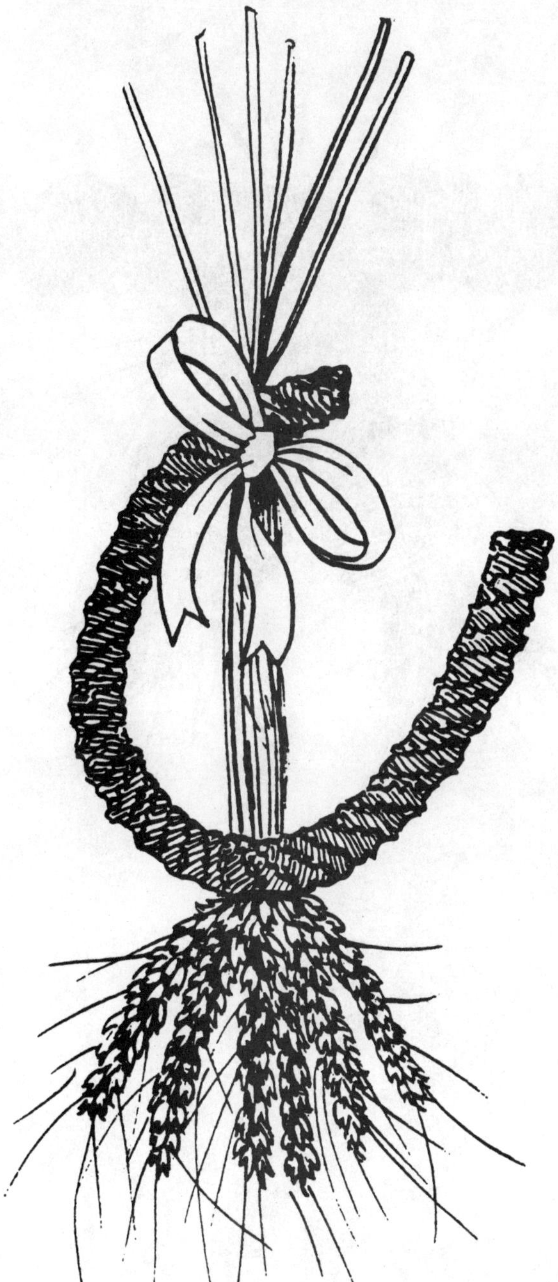

Double Bells And Woven Wheat Core

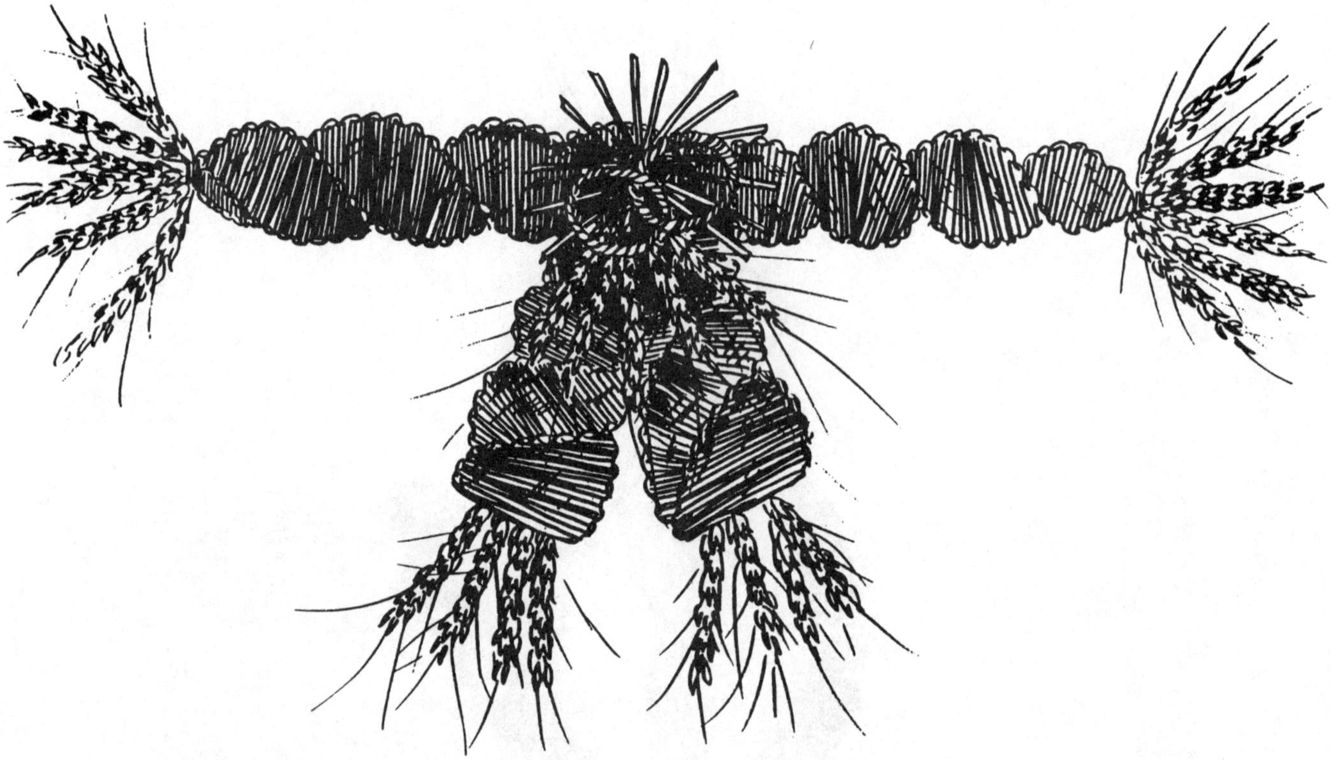

Design With Six Christmas Trees

Oval

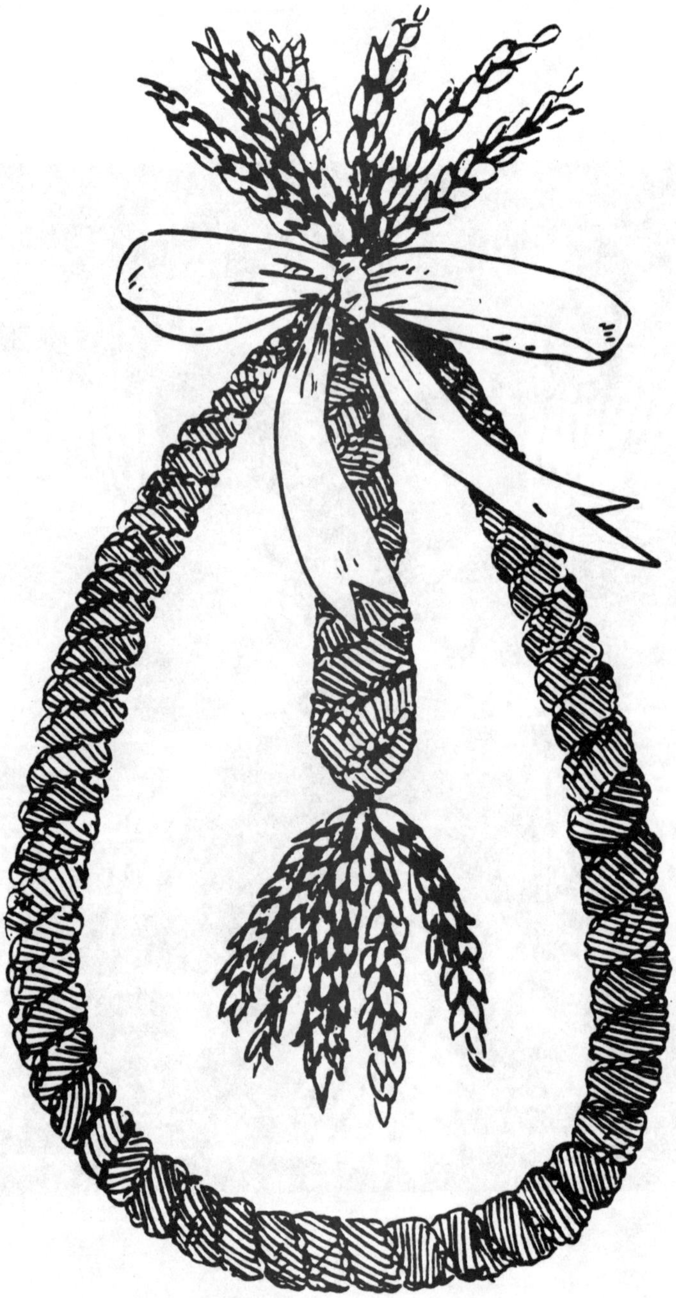

Five Straw Plait

Interwoven

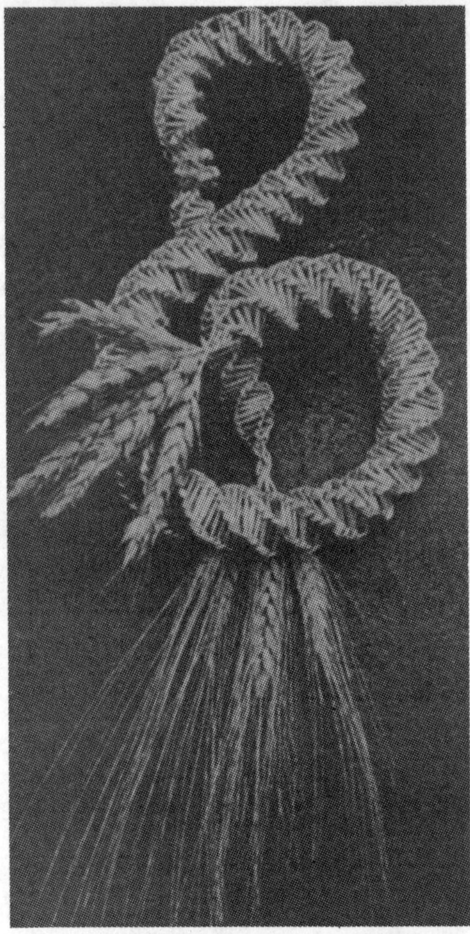

Wheat Dolly Wall Hanging

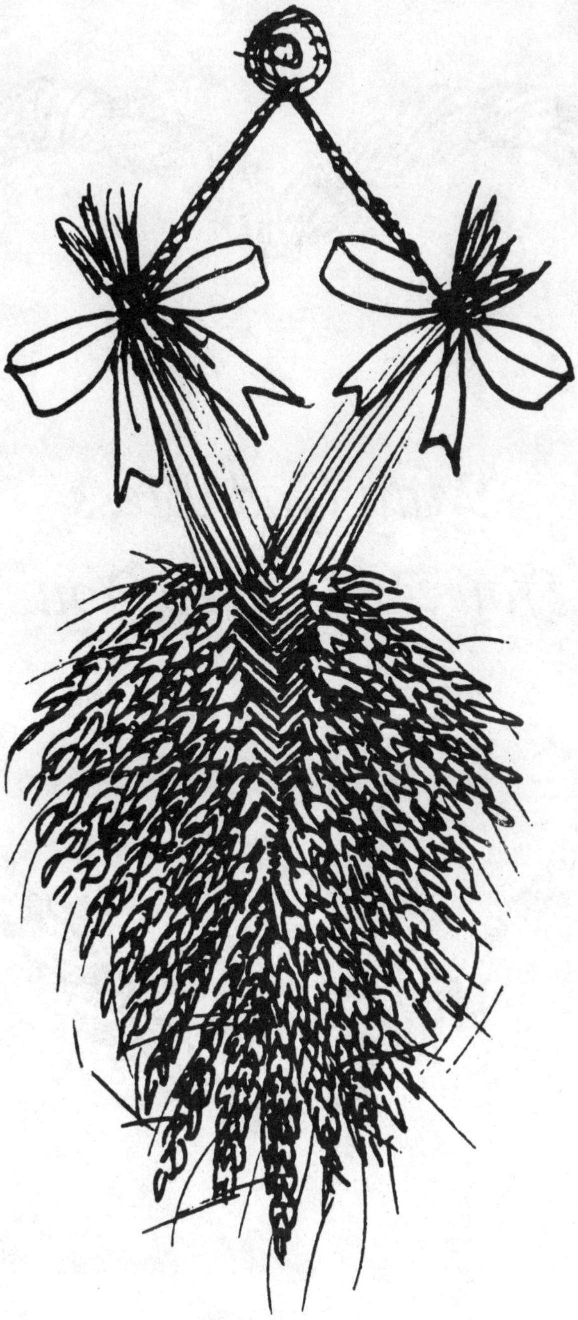

"May God Bless
You And Keep You"